The Post

A "Nelson's Men" Series about the Korean War

Book One

By
Dennis Kennelly

The Post 2nd Edition
Copyright © Revision 2020 by Dennis Kennelly
ISBN 978-0-9971530-8-8
Library of Congress: 2016921215

Cover Image: Shutterstock

La Maison Publishing, Inc.
Vero Beach, Florida
The Hibiscus City
www.lamaisonpublishing.com

From the Author

Although a work of fiction, this story reflects the time and attitudes of world events. Actual time frames and depictions of noted military and political leaders are facts. My story and its characters are a product of my imagination.

My daughter, Christine, inspired me to begin this new journey. "Anything is possible! You just have to try."

Peg, my wife, is a steadfast companion, always encouraging, especially through my doubts, often offering suggestions when I hit a wall. I couldn't have done this without her help.

I've endeavored to be faithful to my battle-tested friends. My observations, beliefs, studies, and first-hand encounters with men who have fought in combat, encouraged me to write about what they experienced fighting for America.

"Loyalty to the Nation all the time.

Loyalty to the Government, when it deserves it."

Mark Twain

Prologue

Korea was the most conquered country throughout the centuries. In recent history, by the Chinese, then the Russians, and finally the Japanese, when they defeated the Russians in 1905 and took over ruling Korea until Japan was defeated in 1945. Although Korea saw no battlefield destruction during World War 11, it did not escape the harshness of war nor the brutality of its Japanese rulers.

Citizens by the millions were brought to Japan and sent throughout the empire to work as slaves under the most deplorable of conditions. Militarily organized, the country was obedient and efficient. Besides the human contribution, it supplied food and minerals for the Japanese military machine.

During forty years of occupation, the Japanese did a few good things for the county's infrastructure. The only two potential ocean ports, Pusan and Inchon, were built into thriving centers during this time. Where none existed, main roads and railroads were constructed along both coasts, from the southern tip to the northern borders. Military ambition dictated these projects, as Japan's quest to conquer China knew no bounds.

Small and extensive mountain ranges dominate the country's geography; rugged terrain exists beyond the few main roads. The climate is heavily influenced by the prevailing winds in the winter from the northwest out of Siberia, which brings terrible frigid temperatures for some years. In the summer, the winds bring rain and heat in the valleys from the southeast coming from the Philippine Sea.

Although the peninsula averaged only about one hundred and fifty miles wide, it could take days to travel across because of the terrain. No river or road ran coast to coast. From the southern tip to its northern border, the country spanned roughly eight hundred miles.

America didn't think much of Korea. To appease Stalin, thinking he might cause trouble in Europe, it seeded the northern half of the country to Russia for its two-week war effort against the Japanese. Both halves became Protectorates. Supervised elections were held in 1948. In the north, the Communist Democratic People's Republic of Korea came into power, and in the south, the Republic of Korea was born. A year later, as agreed to by treaty, both the Russian and American Occupation forces withdrew. Both countries left military advisers and trainers behind.

As 1950 unfolded, the mood in America was light-hearted. War was a bad memory that was rapidly fading. The American Military wasn't so happy, though. The once mighty and proud were no longer. Five years of peace had savaged the resources of every US service arm. Peacetime regulations drastically cut manpower, blocked promotions, and enforced demotions. The government's military budget had scant resources for anything new or deemed replacement. *Make do*, was a familiar lament throughout the ranks.

Signs of impending danger loomed large. Many chose not to look. Some decided a deaf ear was safe. Careers were at stake. Look, but don't see; listen, but don't hear! The country's geography was challenging, where mountain ranges dominated the terrain. The climate was either moderate and wet or frigid and bleak. It was a harsh land.

This is a story about Colonel Nelson, a man who did see the danger signs and who acted, but mostly, it's about his men.

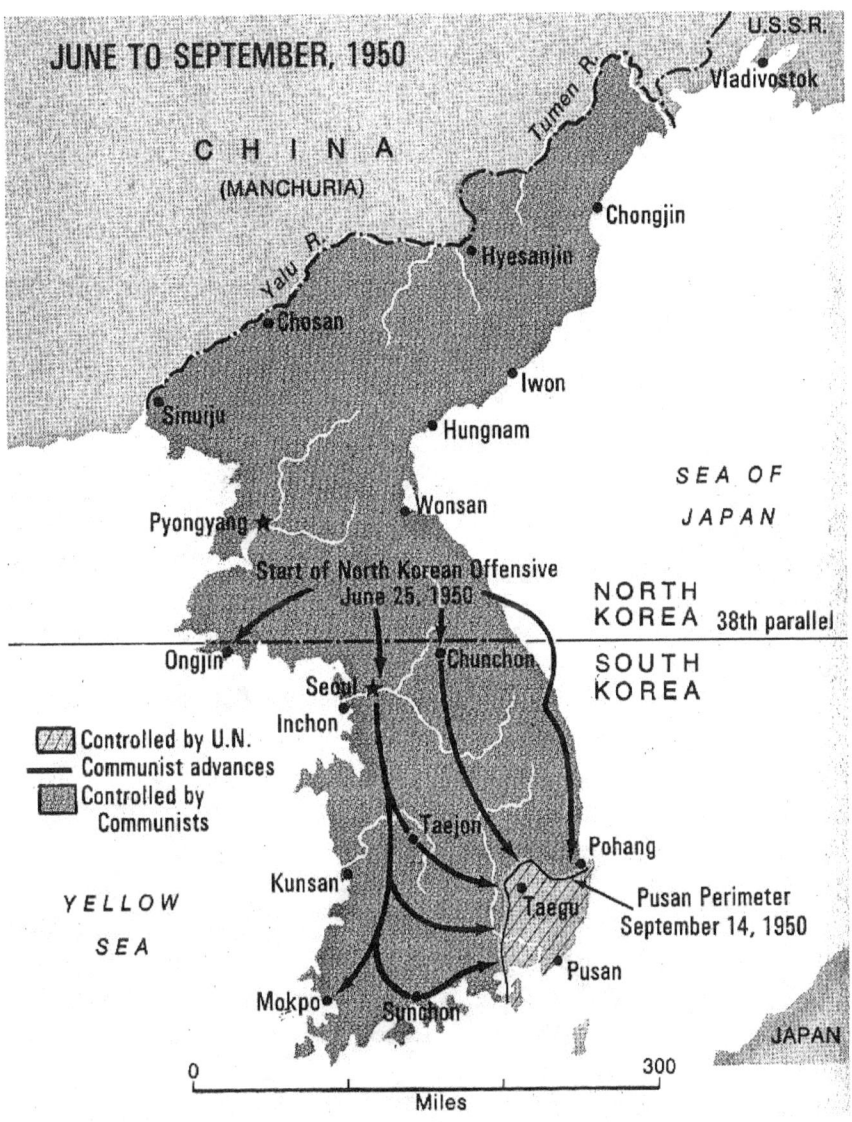

JUNE TO SEPTEMBER, 1950

U.S.S.R.

Vladivostok

C H I N A
(MANCHURIA)

Tumen R.

Chongjin

Yalu R.

Hyesanjin

Chosan

Iwon

Sinuiju

Hungnam

SEA OF
JAPAN

Pyongyang ★

Wonsan

Start of North Korean Offensive
June 25, 1950

NORTH
KOREA 38th parallel

Ongjin

Chunchon

SOUTH
KOREA

Seoul ★

Inchon

▨ Controlled by U.N.
— Communist advances
▨ Controlled by
Communists

Taejon

Pohang

Kunsan

Taegu

YELLOW

SEA

Pusan Perimeter
September 14, 1950

Pusan

Mokpo

Sunchon

JAPAN

0   300
Miles

Chapter One

June 22, 1950 – Wednesday
Korea – Kimipo Airport – Five miles Northwest of Seoul
– 0830 – (8:30 am)

Billy watched the lone soldier exit the daily C-47 "Red Eye" out of Japan, drop his gear, and do a slow three-sixty.

Looks dangerous, Billy thought, like a hunter surveying his favorite grounds. Best not piss him off. He pulled his truck up next to the plane and got out.

"You, Joe?" He barked.

"Ain't nobody else on this plane! You, Billy?"

He nodded. "Welcome to the land of nothing."

"Well, that explains all the empty travel guides and poor onboard food service. Are you taking me to my posting?"

"No, Hank is. As soon as he can find a jeep that runs. I've got to unload this bird and deliver it today. I don't have any time for a joy ride with a private." Billy said with harshness in his voice.

"Is everybody here as cheerful and friendly as you, or is this you just being you?"

Billy realized he might have gone too far, was going to say something, when the jeep showed up and did a quick stop near them.

"About damn time, Hank! Get this private up to The Post and get back here before noon!"

"You need to fix the other two jeeps, Billy, 'cause this one doesn't sound so good either. Be back at noon if nothing holds me up, but you know, shit happens!" He looked at Joe, "Come on, get in. We're on a schedule!"

Joe threw his gear in the back, turned to the sergeant. "Not sure what your problem is, but you need to figure that out fast because you're looking for trouble from somebody not as nice as me."

Joe got in, and Hank floored it.

Looking at Hank, Joe realized he was just a kid, maybe eighteen or nineteen. Yeah, just like me so long ago, he thought, innocent. Never seen combat, never killed a man, never watched his buddy die.

"Did something bad happen to him last night, or is he always like that?"

"Billy's hard to describe. He's okay when he works alone, just doesn't work well with humans. Loves machines, engines mostly, he can fix just about anything, and really good at it. But people? He doesn't have a clue! He's on edge right now. Our lieutenant got hurt, had to take over. He crazy hates it."

As they cleared the main airport gate, Joe eyed the Korean soldiers at the sandbagged machine gun positions on either side.

"You don't seem to have much staff here. How long you been here?"

"Eleven months, eight days. Going home in twenty-two days. Yeah, not much of anything here. About the time I got here, the Army was pulling out; the staff kept dwindling until it's just the LT, Billy, me and Arnie now. A few pilots come and go, but they keep to themselves. We get one, two planes a day, sometimes none, so it's not hard work. Taking stuff or personnel out from the base is my favorite thing to do, so today is a good day for me."

"How many American troops here?"

"Not many. Seoul has a few buildings with a lot of officers. I make deliveries there maybe once a week. Other than that, just a few small training camps outside the city."

"Know what outfits they're with?"

"No. Things I deliver to Seoul are all addressed to KMAG headquarters, but I don't know what that is."

"Me neither."

After driving through the capital they entered into its northern suburbs, Joe sat back and took in his new country. Impressive, he thought, there are no signs of war, unlike Japan. I really should know more about this place. Shit, I don't even know where I'm going.

As they drove on, he saw a lot of military activity. Jeeps and trucks, checkpoints, and small units patrolling.

"Pull over, Hank! I need something."

"What's up?"

"Not feeling comfortable."

Joe got out, reached back into his gear, pulled out his combat belt, his .45 automatic pistol in its shoulder holster, then grabbed his Thompson submachine gun from the rear seat and put it in the front. He buckled up, strapped on his holster, got back in, and repositioned his Thompson across his lap, at the ready.

"Feel better now?"

"Yeah, let's go. How far?"

"Just south of the border, about forty miles. It's been a few months since I've been there. Usually, it's an easy ride, but there's a lot more activity today than I'm used to seeing."

"So how is it, the Post I mean?"

"I don't know anything about it except seeing the entrance gate and a cement building on a hill. The whole place is like a fortress, with multiple layers of barbed wire surrounding it. That's as close as I've ever been. Only been there twice before. Billy's been inside but won't talk about it. Says it's all secret, you know, like he's a big shot, and I'm a piece of crap."

"Why do we see so much military activity?"

"Our LT tells us that the Korean government is nervous, and is responding to communist infiltrators from the North that have been making a lot of trouble, blowing up things, and assassinating local political leaders. Recently, it has gotten pretty tense. We've had flare-ups before, but this seems really serious."

3

Twenty miles out, traffic thinned, and the landscape changed to rolling rocky hill formations, with high brown grass just off the road. When they crested a hill, looking east, Joe could see a valley of dry terraced rice paddies flowing toward distant, mostly grayish-brown mountains. Occasional rooftops dotted the landscape. Small dirt paths angled off from the road, sometimes toward houses. Mostly they just disappeared. A rare run-down hut near the road was a stark reminder of the condition of most Koreans.

"Is the weather always this wonderful?"

"Spring and early fall are pretty nice. Summer is about to start soon when the monsoon season rolls in; then it gets hot and sticky, and rains almost every day. Winter just sucks!"

"Damn! Our luck to get behind a ROK truck." Clouds of dust from the dry unpaved road engulfed them. He was hogging the highway, and unless the truck moved over, they couldn't pass it on the narrow road.

"What's an ROK?"

"You **are in** Korea. It's their Army. You know… The Republic of Korea. That's what we call all things South Korean Army." He grinned. "You need to catch up on where you are."

"Right! I kinda left in a hurry."

The truck in front of them exploded.

Hank yanked the wheel to the right out of instinct. The jeep hit the grass rise going thirty as he slammed on the brake.

On impact, Joe ejected out of his seat, hit the windshield edge, tumbled into the grass like a parachutist on landing and rolled, finally becoming prone with his Thomson, ready to fire. He couldn't believe he was alive, not a scratch.

While he was flying through the air, he heard bullets hitting the jeep, shattering glass and parts. The sounds were piercing.

Laying still, he heard voices yelling over the automatic fire, and they were getting closer. Joe jumped up and started shooting, immediately killing two men dressed in civilian garb, carrying Russian made sub-machine guns.

Joe then ducked, rolled to the left, and got still. He heard more voices on the other side of the road. The shooting had stopped. In a low crouch, he moved fast toward the burning truck, his head just below the high grass. At its front, he stopped for a second, popped his head up, didn't see anything, then, partially hidden by the smoke, he ran across the road. Focusing on the last position where he thought he heard voices, he knelt and listened, heard only the crackle of fire from the burning truck, nothing else. Lifting his head, he scanned the area and saw a little movement of grass thirty feet to his right. They're crawling away, he thought.

He stood and fired a long burst, then ducked and rolled right at the same time slamming in a new magazine and set the first round. Crawling another three feet, he stood and fired another long burst in the broader arc. Reloading, he approached the area. Two more dead civilians, weapons visible. This was it, he thought, four guys on an ambush mission. Must have planted a big mine for the truck. It should've been easy for them, but they sure didn't count on a turn of bad luck. Me!

Already knowing what he'd find back at the jeep, it burned him when he saw the young soldier slumped over in his seat. His mind kicked in and started a conversation. Twenty-two days left became an eternity for you, Hank. I'm sorry I couldn't save you.

What the hell have I gotten myself into? Not here two hours and shit's finding me. Maybe I should've stayed and been court-martialed. At least I'd be... What? Safe? Hey! Wake up, asshole! Get moving, check the jeep. Can it run?

Taking a quick look around the jeep, he saw no damage to the engine or tires. Looking at the young kids' bloody body, it brought back an unpleasant flash memory that made him stop.

"Damn it!" Joe said out loud. "It's been five years since I've seen the ugly side of life. So, old boy, I think it's time you face the fact that this new posting might get horrible." Still talking aloud. "Okay, Hank, I'm gonna move you into the back, buddy. Need to get this jeep moving."

5

Dennis Kennelly

He pulled out onto the road and moved north. "Jesus," he said aloud, "I don't even know where this damn place is, Hank. Wish you'd been more specific, kid. I hope this barbed wire fence gives me a clue."

Chapter Two

June 22, 1950 – Wednesday
Korea - The Post – 1050 – (10:50 am)

Arron was on security, patrolling along the trench outside the building. He was expecting the new guy, so he stayed on the east side facing the road.

He saw the jeep stop at the gate, lifted his binoculars, "Oh, crap!" He ran into the building and yelled, "Rusty! Jeep at the gate, driver only. There's a body in the back!"

"Oh, Christ! Arron, get'em up here!"

Rusty got to the jeep as it pulled in next to the building. He saw the shattered windshield and some bullet holes, then focused on the body in the back. One side of the soldier's head was gone. He recognized him from trips to Kimipo but couldn't remember his name. The memory lapse must have shown on his face.

"That's Hank," said Joe, "I'm Private Joe Cappanela."

Rusty turned to Joe. "Glad you made it, Joe. I'm Rusty Fabrino." He extended his hand, and they shook. "Looks like you had a tough ride. Follow me."

At his office doorway, Rusty called over to a guy sitting at a makeshift desk across the room. "Hey, Larry, would you get us some coffee?"

"Sit down, Joe. We don't have any formalities here, relax, and take off your combat gear."

As Joe laid his waist belt on the floor, Larry entered and set two steaming mugs on the desk.

"Thanks. Larry, meet our new guy, Private Joe Cappanela, Sergeant Larry Sanders." Larry extended his hand, and they shook. Then Larry left.

"My orders say to report to the Post commander. Who's that?"

They both reached for a mug, took a big gulp. Joe's face brightened with the taste of the coffee.

"Hey, that's good Java."

"Yeah, it is. We have Arron to thank for that. That's the feller that met you at the gate. He does something to coffee, and pretty much anything he cooks makes it unique. He's got a gift." He took another gulp then continued.

"I'm the Post commander, Joe."

"But you're a First Sergeant! You don't have an officer?"

"Correct. You'll find that all the men here are sergeants or first sergeants, but I'm first among equals."

Rusty leaned back in his chair. "Bitch of a first day. Tell me what happened."

Joe recounted the events of the ambush.

When finished, Rusty said, "So, you got four guys who had a dead drop on you. I gotta say, you're even better than what your file says."

Joe looked at him, tilted his head like he wasn't sure of something.

"How'd you know what's in my file? I haven't given it to you yet."

Rusty nodded. "I know you don't have a clue about what's going on here, so I'd better explain."

Joe leaned back. "You have my undivided attention."

"I've had a copy of your file for over two months. You were our number one selection to replace our backup radio guy should something happen to him. We have a backup man ready for everyone here. It just happened that Marvin got sick, and you screwed up at the same time, so it made it easy to pull you out and transferred without much of a fuss."

8

Joe moved to the edge of his chair, annoyed. "What the hell are you talking about? What is this place? Who is "we"? Who is the real boss, and what outfit am I actually in?"

"Settle down! I'll answer every question you have. You may not believe this, but every man here is the elite of the elite, experienced combat vets, and all volunteers. You're the only exception. Well," Rusty grinned, "that's not entirely accurate, is it? You had a choice; Leavenworth or here. So I guess you really did volunteer."

Joe chuckled, liked his talking straight, and something else that he couldn't put his finger on. He took another sip of coffee.

"Okay, Rusty, maybe you better start at the beginning."

"It's complicated, but I'll hit the high points. Almost two years ago, when the United States was pulling the occupation forces out and turning everything over to the South Korean newly elected government, a few people started worrying about the Communist North." He paused a moment.

"Our boss is Colonel Jim Nelson. He's G-2 (Intelligence Chief) of the Eighth Army and the brainchild of this Post and three others. Back then, he figured that sometime soon, the Commies wouldn't settle for only half of Korea and would make a play for the whole peninsula. He felt an early warning system could forestall an invasion or even prevent it. Knowing that South Korea didn't have the expertise, and America's political powers that be wouldn't want to be involved in a potential hot-potato like spying in and on newly independent countries, a secret plan was hatched.

Joe now knew he was sent to the right place.

"Technology enters here, as does the CIA. As a backdrop, Colonel Nelson goes back to the OSS days with Donavan. They became friends after a few special assignments by our colonel when he was a major."

"This Colonel Nelson is with the CIA?"

"No, no. Sorry, if I'm confusing you. He just has close ties with them. We work with them closely now on intelligence gathering, using their secret equipment. CIA worked with Eighth Army and developed a radical new way of listening to radio frequencies, capturing them and then with another new technology, transmitting these without wires over very long distances."

"I think I know something about the intercepting equipment."

"Yeah, I think you do. We'll get to that in a minute. Anyway, these two innovations clinched the colonel's idea of establishing secret listening Posts along the 38th parallel. The CIA supplied the equipment and set up a decoding operation."

Rusty stopped for a moment then said, "Joe, this is all Top Secret stuff I'm telling you."

"I don't suppose my Top Secret clearance has anything to do with my being here?"

"I'd like to say it was just because you're special, which you are, but yeah, your expertise in radio communication and your work on this secret monitoring equipment were decisive factors in your selection. Which, by the way, begs the question. Why did you leave your part-time duty at the research lab? I couldn't figure that out."

"Why? Because that's when things started going downhill."

"What do you mean?"

"I made first sergeant about six months before that. There wasn't much to do with my platoon, except mostly paperwork. We weren't authorized to train, except to march for an hour a day, and the rifle range was available every other week so that we could fire one clip or one magazine. That's why I got into the Advanced Radio program and then got involved with this new secret stuff. I loved it." Then he frowned.

"But then my company got a replacement second lieutenant, fresh out of the Academy. It was water and oil between us from the get-go. To make matters worse, we had gotten a new company captain before that, and he knew my new LT. They were friends. Anyway, this LT wanted my full-time attention on the platoon, not to do anything special, but to just be around. He believed that spit and polish would somehow win future wars. I also think he was jealous or something about my involvement with the secretive people at the Lab. Everybody assumed the civilians were CIA. I never knew before, but now I do."

He finished his coffee.

"Anyhow, he set me up, got to the captain, who then ordered me to quit the lab. That pissed me off something awful. A slow fire built inside me, then erupted when this second LT laced into me and dared to humiliate me out in front of my whole platoon because I suggested a different approach to a tactical infantry situation we were studying. But then, when he said, "Shut the fuck up" and that I was "stupid," and that he was from "The Point," well, that's when I knocked him out."

Rusty laughed. "I knew about the knock-out part. But leaving the lab, thanks for clearing that up. This LT sounds like a real jackass."

Joe nodded. "He was, but I should've controlled myself. I just lost it."

"So that you know, your colonel is a friend of my colonel. He called Colonel Nelson when this went down, because he knew of his interest in you. Fate intervened when Marvin got sick the day before. Oh, yeah. Do you know what your colonel said about this incident?"

"Not a clue."

"He told Colonel Nelson that you should've broken the little shits' jaw."

Joe laughed. "I should have!"

"Yep, agreed. Now, where was I?"

Suddenly, a huge guy appeared in the doorway. "Hey Rusty, Larry's got Colonel Nelson on the cable phone for you."

"Okay, sorry Joe, I've got to take this. We'll pick it up later." He gestured to the big man. "Come on in Red. I want you to meet Joe, show him around. I'm gonna be a while. Joe, this is First Sergeant Kenny, "Red" to all. He really runs this place, so what he says goes."

Joe stood, turned to Red, and shook. He was looking at a force of nature. At 6' 2", Red was 200 lbs. of solid muscle, red hair, and intense green eyes. Handsome in a hard way, a square jaw-line sported an impressive four-inch scar. His presence was intimidating.

As Rusty walked out, Red said, "You passed the first test, so welcome to our little club."

"What test?"

"Being fast and lucky. You survived that ambush, Private." Red smiled.

"Come on, let me show you around. Outside first, follow me."

They walked out the only door in the building. It was steel, opened out, immediately faced a concrete blast wall; one could turn left or right and either go to the vehicle parking area or down ramps into the four-foot trench line that snaked around the entire building. Joe followed Red down into the trench.

"Jesus, Red, this is impressive! You dig this trench by hand?"

Red laughed, "We'd still be digging it if we had to do it by hand. This building is sixty by one hundred, and you haven't seen the corner forts yet. No, we got help from a South Korean unit stationed up at the border when they were fortifying their positions. They lent us some heavy equipment. We made good use of it. You'll see."

As they got to the northeast corner fort, Joe didn't understand what it was. He just saw a half-circle jutting out of the corner of the trench that had a rear entrance wrap, much like a rat maze, designed to protect the occupants from rear explosions. But when Joe entered, he understood what an extraordinary position it was. Immediately facing a mounted thirty-caliber machine gun in its center, the view beyond scanned out over the sandbags that wrapped the fort to see a vast open field below. Moving to the front edge, he now saw the sloped thirty-foot elevation as it ended in a flat area. Looking farther out, he saw the barbed wire rings, three. Evenly spaced apart, with the most distant ending near a tree line, maybe four hundred yards away.

Red stood behind him, smiling at Joe's back, knowing what he was thinking. "Not bad, heh." He said casually.

Joe turned with a big grin. "There's more, isn't there?"

Nodding. "Oh yeah!" Red's smile was like a cat with a canary in its paws. "Mines! Lots of mines. I don't think you could walk six feet in any direction without getting killed."

Joe did a quick look back at the vastness of the field, shook his head. "Jesus! Gotta be a lot of mines."

"As best as I can figure, about thirty thousand."

"What?!" Joe was stunned.

"They're mostly anti-personnel, different types. From surplus captured German stuff to our own mixed types. Got about five thousand anti-tank mines mixed in."

"Where in the hell did you get all of them?"

Red showed how proud he was of his efforts and grinned. "We couldn't go through regular channels because, well, because nobody knows about this operation. It's beyond Top Secret, especially from our own Army. So we had to improvise."

Joe squinted his eyes, realizing that he, and this Post, probably had a very short lifeline.

"Sometimes the Army outdoes itself. Like last year, when the Commies ran Chiang Kai-shek and his army out of China to Formosa, we thought they might invade the island. The Army sent so many mines and barbed wire to Formosa that, after a while, the ships diverted to Japan for fear that the sheer weight of all that stuff might sink the island. So we have millions of mines and miles of barbed wire stored in Japan. Between Colonel Nelson's pull and my connections in Japan with other sergeants I served with, the trucks began arriving within a few weeks, no paperwork, and no trouble. We started planting the mines and stringing wire immediately."

"Amazing!" Joe smiled but was concerned by its implications. "Do these barbed wire lines go all around the entire Post?"

"Yeah, except the entrance road from the gate. That's got special protections."

"Red, you made the Post into a fortress. You think it was necessary?"

"Colonel Nelson believes this area will be the North Korean's main line of attack, so delaying them is a big priority. Also, I'm

13

paranoid. Having been overrun three times in the Bulge by the Nazis, I vowed that if I was ever in charge of setting up a defensive position, it was going to be incredible, a not – getting - through type, which would punish anyone trying to get my men and me. Necessary or not, I'd rather be safe."

"Can't argue with that. Sure makes me feel better."

Red led him around the entire trench and back into the operation area of the building.

"This room has our essential equipment, is manned round the clock. Meet Ray, Sergeant Mathews, our radar guy." They exchanged hi's.

Ray pointed at the radar screen. "You probably saw the dome on the roof when you came in. We can see up the entire valley for fifty miles and to the mountains on either side. The North Koreans don't have too many planes, but they have enough to cause lots of trouble if you don't know they're coming, so we watch."

"Say hi to Sergeant Sanders. Larry's our main radio guy, an explosives expert, wiring genius, and the resident nerd and sometimes jokester. You'll be spending a lot of time with him." Larry's desk was crammed with multiple phones and stacked metal boxes with blinking lights and oscillating glass scopes. It wasn't a typical radio desk. The two nodded. "We met earlier." Said Larry. "Welcome aboard, Joe."

"Larry has some new radio stuff; he'll fill you in later."

Joe's face lit up as he turned to Larry. "Using the Wide Array Spectrum Receiver Collectors?"

Larry gasped. "That's ultra-secret! How do you know about it?"

"Right." Joe nodded and pointed to a corner of Larry's desk, "That box over there with all the lights flashing, well, I've only seen one other like it, and it was in the Secret section of the Advanced Radio training section I worked in for a while. I have clearance, so I spoke with one of the guys working on it. They were trying to increase its range. They also said they had a few in the field. I guessed you are using it. Perfect application. I never did find out its range, so how far will it pick up signals?"

"You're a scary guy, Joe. Not many men would have put that together. So you'll be monitoring it, and it's probably the most useful piece of equipment we have. The range depends a lot on weather conditions. We pick up any radio signals within ten miles. On good days, that can go up to fifteen or so miles. I'll fill you in later."

"Great! Love new stuff."

Larry grinned. "We have some other neat things, too."

"More? Like what?"

"Do you know anything about VLF?"

"What?"

"Very Low Frequency."

"Never heard of it."

"OK, we'll talk later. One out of two ain't bad. And by the way, as you'll discover, I'm only a jokester wannabe. There are professionals in the house, and here comes one right now. Meet Foxy."

Foxy moved in, hand extended. "Hey, welcome to our little slice of heaven. Heard you didn't treat the welcoming committee very nicely."

The guys smiled, and Joe said, "Next time, the committee might bring coffee and cake. Foxy, good to meet you. I've been looking for my little slice for a long time. Hope you don't mind sharing?"

Foxy grinned. "No worries. There's plenty to share. The air is free, and that's pretty much it."

Red chimed in, "Foxy is half Korean and our only interpreter. But he's a lot more than that, pretty good at a lot of things, like training the local ROK units. He's a superb teacher, almost as good as me." He said with a smile. "Also a great planner. And a better night fighter than most. Foxy served in a recon unit for Third Army from Normandy to the end. Out front most of the time, he got a few medals along the way."

Joe looked at Foxy. Most Koreans were small of stature, maybe five-five, but being half-American, he was six foot, with Asian eyes and chiseled features that showcased the best of his Swedish American dad and his Korean mom.

"Don't listen to Red, he tends to embellish. You know how the Irish are; drink too much, talk too much, and they love their stories."

Joe smiled. No egos, no one trying to prove anything, he liked what he'd seen so far. This is some outfit, he thought, maybe I am lucky. I could be in Leavenworth.

Joe asked Foxy, "Do you have any relatives here?"

"I do," said Foxy with a slight change in his demeanor, "My grandma lives in Seoul with my cousin, seventeen now. She has a small restaurant. They're my only family here. They survived the Japanese occupation because the Japs liked her food so much they didn't send them to a work camp. The rest of my family here weren't so lucky. I get down to Seoul a few times a month to visit and get some great home-cooked food. You'll love my grandma's cooking. I'll take you next time. You ever have any real Korean food?"

"Are you kidding? I never even heard of Korea before I got here, but I'm willing to give your grandmas' cooking a shot."

"Good! Next time I'm planning to go, I'll let you know. Tomorrow, Rusty wants me to take you up to the border and meet Captain Hui. You'll like him. He speaks English, plus you'll get to see the terrain. You need to become familiar with the area."

Red shook his head. "Plans have changed, Foxy. We need to bring Hank's body to Kimipo in the morning. I'll talk to Rusty about it and get back with you."

Red picked up Joe's duffel bag. "Come on, I'll show you the rest of the castle."

Foxy called after them, "Hey Red! Don't forget to give him my cheat sheet."

"What's that, Red?"

"Foxy made a page of standard Korean sayings to study, you know, like **Stop or I'll Shoot**. Might come in handy, I'll get you one later."

16

"Yeah, I guess I should be able to say a few words, good idea."

They walked down the hallway and stopped at the first room on the left.

"Your suite, Joe."

It was a dorm-like room with five sets of bunk beds, two on each sidewall, and one straight ahead.

"Your bunk's on the right, top front. It ain't much, but it's home."

Joe turned to Red, "I lived in foxholes for three straight months on Okinawa. This is really fine!"

"Glad you feel that way because the latrine is field issue, next door on the left, two pots in boxes. Field makeshift shower as well. We carry water in by jerry cans that we get from a stream south of the Post."

"I guess we boil our drinking water?"

"Oh, yes, we do." Red went on, "Even in this wilderness. It's a dirty country, lots of people wandering around, shitting anywhere. Then there are the dead bodies in the rivers as well. Crazy stuff, kind of like our Civil War, the south hates the northern Commies, the north hates the capitalist, so the killing goes on. The sad thing about both wars is that the poor and ignorant suffer the most."

"You've got a philosophical streak, Red."

"I guess. Maybe I've just seen too much death. Killing soldiers is one thing, but civilians...?"

Red turned, walked out into the hallway, and pointed to the room across from the dorm.

"That's our kitchen and mess hall. The next room down is our general storage area, including weapons and ammo. Behind that door at the end of the hallway are two generators, fuel, and water storage. There are no exit doors or windows. We run an exhaust fan through a small opening in the side on the south wall."

Red glanced at his watch. "Let's go. Rusty should be off the phone by now."

He knocked on the sidewall of the no door entrance.

17

"Come in," Rusty said and glanced over at Red. "I told Colonel Nelson about the ambush and Hank. He said he'd notify Billy, hoped we could bring his body to the airport in the morning. I said we would so set that up. Also, I need you to check on Arron up in the turret. He had an idea about putting a machine gun up there. See what you think."

"I already know what I think."

As he walked out, Red gave Rusty a thumbs up.

"Joe, please sit." Rusty was standing behind his desk.

Joe noticed the maps on the walls around him for the first time. It was a small room, maybe ten by ten. The desk, like all the others in the OP room, was just wood planks nailed to sawhorses. The chairs were real. Joe wondered how they got them, then thought of Red and his connections back in Japan.

"So, what do you think of your new home?"

"Besides the roadside welcome, I like what I see."

"Any questions so far?"

"I only saw ten sleeping bunks. Is that all the men we have? Are the other Posts the same?"

"I have ten men, including you and me. The other posts have the same composition of men and ranks except for the one to our east. It has a captain, a Captain Jordan, kind of the commander of all four posts."

"What do you mean by *kind-of*?"

"I have a history with Colonel Nelson. He was my battalion's G-2 when we landed in Normandy. I did a few special operations for him. We got close. Then he got kicked up to regiment. By war's end, he was with General Walker as Corps G-2."

Rusty sat down on the edge of his desk and continued.

"When he put this special Korea thing together, he wanted me here, at this Post, because Colonel Nelson trusted me to carry on when the fight came, as he knew it would. The others are called Post #1, Post #3, and Post #4, and we all have the same secret CIA spying equipment.

"But this Post is special. We labored in secret for almost two years to harden this site to withstand a prolonged attack for at least two days, maybe more. So we're meant to slow the enemy on this main invasion route for just a bit, and then go on to continue fighting as guerrillas, and helping the South Koreans. I hope Red talked to you about this."

"He gave me a great tour, First Sergeant, but he didn't elaborate upon our mission, or about what happens after the rockets fly."

"Joe, please stop calling me by my rank. I consider you my equal. You were screwed by a peacetime system that favors connected officers. First names here, please, I'm Rusty, you're Joe, and, oh, although you're officially a Private, we're working on getting your First Sergeant stripes back. And oh, all the men here know how you ended up here. So you know, if they didn't approve, you'd not be here."

Rusty went on.

"Beyond our basic roll of spying, we are all tasked to help our South Korean fellow soldiers in any way possible. And, as I said, should an attack from the North occur, we are to engage, slow them down, and disrupt the enemy in any way possible. Colonel Nelson thought I was the best man for the job here. He couldn't promote me because of the *peacetime* regulations imposed on the Army, so he explained the situation to Captain Jordan. Jordan was good with it, and so was I. Captain Jordan's a very competent officer but just not what Colonel Nelson wanted for this Post."

Joe scratched his head. "Never heard of such a thing. Colonel Nelson must be special. Come to think about it, so must you and this Captain Jordan. Did you secretly defeat Germany or something?"

Rusty laughed at that.

"Well, Colonel Nelson is unusual. He doesn't think like a typical officer. And no, I'm nothing special, let's just say that Colonel Nelson thinks I'm exceptionally resourceful and leave it at that."

"Fair enough. I'll judge for myself."

19

"I like that about you."

"Well, having been surrounded by idiots a few times, it's the only reason I'm still breathing."

Rusty smiled. He knew Joe wasn't kidding. He recalled reading about his unit's experience fighting on Okinawa. Three solid months on the frontline, *in-your-face* fighting, with 80% causalities. He'd had four replacement officers during this stretch. A real tribute to his survival skills, he thought, and to his good luck.

"What's the current situation?"

Rusty frowned.

"Things have picked up recently. About a month ago, activity on the highway flared up with a few ambushes on ROK vehicles like today, with you. But a week ago, local action started getting interesting. Radio traffic along the border increased. More North Korean forces began showing up. There were some raids across the border, nothing serious, just light casualties. But these developments trouble me.

"The way I figure it, they're testing South Korean defenses and gathering intelligence. These things are occurring not just here but along the entire border. Two days ago, new North Korean units started setting up very close to the border.

"In response to these aggressive moves by the North, a few new South Korean companies were brought in to reinforce the border defenses in our area yesterday. Things are getting hot here. Our headquarters in Japan doesn't think it's anything to worry about, but Colonel Nelson thinks differently, and I agree with him. Something's up."

"What's the plan if they attack?"

"We get fast, or we get dead. Fast in determining if this is the real deal. Fast in getting the info out and fast in getting the hell out of here if we have to. We have four jeeps and only one road south. This hill we're on is a defensible position, but only if we had more troops. We set this position up so we wouldn't be surprised. With ten men and four machine guns, we barely cover the perimeter.

We'd give'em a hell of a fight. Delay them maybe a day. But with no hope for reinforcement, we'd fold pretty quickly."

Joe was skeptical. "What then?"

"When we escape, we'd go into guerrilla mode."

Rusty heard a familiar voice outside his office.

"Hey, Tracker!" He yelled, "Come in here."

A short, slightly built man with a swarthy complexion and eyes as black as his hair, walked in.

"How!" The soldier smiled, raising his right-hand, Indian fashion.

Rusty and Joe laughed.

"Joe, this is Tracker, our real-life American Indian." The men shook hands.

"If Rusty picked you, you're fine with me. Welcome to our crazy little tribe." He grinned.

"Did you find any activity around the perimeter?" Rusty asked.

"Oh yeah, just like yesterday on the southeast perimeter, outside the first wire. Found the same three-man team tracks on the northeast wire. I don't like it. Something stinks in Denmark."

Rusty couldn't help but laugh. "Only you would quote Shakespeare at a time like this. I don't like it either. Maybe we should set up a little surprise for them on their next visit?"

"Just what I was thinking. I'll get with Red and come up with a plan. Set it up for tonight."

"Let me know the details."

Tracker bowed, smiled, and said as he walked out, "We Indians know our place, General White Face."

"I love Tracker. He's a Chippewa Indian from Minnesota. Been tracking game and people since he was a kid. Even the local cops tapped him to hunt bad guys in the wild when he was twelve."

Joe couldn't resist. "The Lone Ranger was nothing without Tonto."

They laughed.

21

Then Rusty filled him in on security rules, their daily routines. They agreed to meet later with Larry at the radio station to fill him in further.

Chapter Three

Wednesday, June 21, 1950
Japan – Eighth Army Headquarters - General Walker's
Conference Room 1300 – (2:00 pm)

ENEMY APPROACH ROUTES *through Uijongbu Corridor.*

"Colonel Nelson, brief us on the latest developments in Korea."

General Walker and his Executive Officer, Colonel Harris, stood in front of a large wall map of Korea.

Nelson stood facing them; a three-foot pointer stick at his side.

"Things are picking up across all four of the major attack routes." He tapped on the Ongjin Peninsula in the far west. "This is where the North Koreans moved at least one division into yesterday, about five miles from the border. Here, about ten miles to the east, in the Kaesong area, near the entrance to the *Uijongbu Corridor*, we think at least two divisions are setting up about four miles north of the border. Over here, twenty miles east, near Cheorwon, we detected some activity, but we don't know any details. And finally, here, on the east coast, north of Yongyang, we've confirmed that two divisions have moved in."

Colonel Nelson turned to the two officers.

"The North Koreans are going to attack, and soon. I'm thinking probably in three or four days, max, a week. They've moved at least seven out of the ten divisions they have into jumping-off positions. We have good Intel that they'll be moving in artillery and tanks soon."

"How do you know so much about the North Korean's force movements?" Asked Colonel Harris.

"I'll get into that in a minute. Let's just say that we're using experimental top secret electronic equipment to monitor radio traffic near the border."

General Walker nodded his go-ahead to Nelson, and he continued.

"We believe the main attack force will concentrate in the Kaesong area. I think we are underestimating the present NK unit strength there. It's a straight shot down this valley plateau to Seoul. About forty miles. They could link up with their forces from the Ongjin area after that spear breaks out from across the Hantan River. They'll have a formidable attack force to take the capital."

"What's the ROK doing?" General Walker asked.

24

"The military is taking this situation seriously but don't think an invasion is imminent. The North Koreans are claiming they're conducting maneuvers, and Seoul believes them. They haven't canceled leaves, nor put their troops on full alert. But they have moved some forces closer to the border."

"Do we have specific ROK troop movements?" Asked General Walker.

"Yes, sir. They've moved the 17th Regiment of the Capital Guard Division into the Ongjin area, about five miles south of the border as a reserve for one regiment already there. In the Kaesong area, the 12th Regiment of the 1st Division moved closer to the border. The 11th and 13th Regiments are in reserve just south of Kaesong. The 2nd Infantry Division was moved in as a reserve for the 1st Division about ten miles further south of Kaesong. On the east coast, an additional regiment repositioned to reinforce the 8th Division in place there. That's all we know at this time."

Colonel Harris walked up to the map. "Do we have any US troops near the border?"

Nelson responded, "Colonel, we probably have less than six hundred men in the whole country, mostly KMAG, our military advisory group. Of those, probably two hundred are advisors, and the rest are staff. Few are near the border except for my men."

"Please elaborate, Colonel." Asked Colonel Harris.

Colonel Nelson frowned. "We have four clandestine listening posts near these invasion routes, all using secret electronic monitoring equipment. Their primary mission is to do early warning and gather intelligence. Their secondary mission is to engage in special operations that, in the event of an attack, will disrupt the enemy."

"Colonel!" Colonel Harris interrupted. "How many men are stationed at these Posts, and can we defend any of them?"

"They're all ten-man teams, Colonel. All are handpicked, and each man has a specialty. Most are non-coms, and all are experienced vets. Think of them as *Spartans* behind the lines, on the front and the flank. They have the capability of being our leading

25

presence in our efforts to harass and slow the enemy until we can bring enough force to stop and defeat them. Right now, they're all we got.

"As to defending the Posts, well, we built one at the valley entrance, in the Kaesong region, I spoke about earlier. It's where we expect the enemy's main line of attack to occur. This Post is heavily fortified, so, best guess, it could hold out for a day or two against a determined attack, but if reinforced, it might fight considerably longer."

"Wait a minute!" Colonel Harris looked confused. "Where did these Posts come from? Who authorized them, and what is this secret monitoring equipment?"

General Walker answered.

"About a year ago, I authorized Colonel Nelson to proceed with this program he designed. Very few people know about this. Sorry, but you didn't need to know this then. The equipment is still secret and will remain so."

Colonel Harris looked annoyed. "So we have forty men at the border, but only ten that will stay and fight at the border?"

General Walker looked over at Colonel Harris and said in a tone only generals can. "Colonel, I hope I misinterpreted your implication. If I have, I apologize. But I, for one, am grateful that we have ANY men at the border willing to help delay the enemy."

"Sorry, General. That came out wrong. I'm just frustrated we don't have more support."

"We're all frustrated, Colonel."

"Colonel Nelson, continue. Any recommendations?"

"General, I'd recommend we push the emergency button and start sending troops."

"Point well taken. Since I can't order a platoon into Korea without General MacArthur's authorization, which he won't give me, do you have any other thoughts or suggestions?"

"I'm as exasperated as you both are. For the last few weeks, I've sent multiple reports about the situation to MacArthur's Chief of Staff, General Almond. I can't get to General MacArthur directly.

General Almond and General MacArthur just don't believe anything's going to happen in Korea. They're focused on Formosa. Of course, politically speaking, President Truman doesn't think we'll ever fight a land war again. He doesn't want to hear anything about a possible invasion by North Korea."

General Walker concurred. "Our President thinks the Atomic Bomb is the end to all wars. What nonsense! By the way, I'm glad I'm not the only one who can't talk to MacArthur. I just get the deaf ear of General Almond. His standing order of no additional troops or resources for Korea is ironclad. They buy the BS our military advisors have told them about the readiness of the ROK forces. They actually think the South Korean Army can easily handle anything at the border."

The General was on a roll. "Even if they're reasonably trained, they're green. They have no heavy weapons, no tanks, nothing to stop those T-34 tanks the Soviets gave the North. No air force! Oh, boy, I do go on."

Colonel Nelson weighed in.

"General, I suggest we ready our forces here in Japan so that our response time will be swift. If we have anything that might help our Posts, we need to send it by air, now. Hopefully, it won't garner any attention from General Almond."

"I think that's the only thing we can do now," The General said. "Colonel Harris, how are our contingency plans going for deployment?"

"I wish I had better news to report, sir. As you know, our Defense Chief has done the bidding of our President with great zeal. He's cut the military to the bone. In some cases, he's removed the bone."

The general huffed, "He sure the heck has!" The two colonel's grunted.

"We're down to 60% strength in all units. Our tanks and most of our artillery are in the States. All Eighth Army weapons are reissued war surplus. We have nothing new. Even our communication equipment is old stuff and works as well as it did in the War, which, as you know, wasn't very good."

He continued, "But that's only half the problem."

"Oh, go on! I can't wait! I thought that was our only problem." Disheartened, the general sat down.

"Sorry, General."

The general motioned for him to continue.

"Transport's our principal problem, sir. General MacArthur controls air and ship resources, and he's not budging. We have few resources here in Japan. Even if we were ready right now, we could only bring in limited forces."

"Give me a number, Colonial."

"General, right now we could maybe fly in five hundred fully equipped troops, per day."

"Holy crap!" General Walker rarely cursed but wanted to just then.

"Until we can get more aircraft in from the States, we don't have much of anything in the Pacific. Same goes for sealift. We don't have a troopship this side of Hawaii. The Navy has a few warships nearby but not much for immediate help."

"That's as grim a picture as I've ever heard, but appreciate you being frank." He stood, started to pace.

Colonel Harris added, "On a more upbeat note, our readiness is progressing. The 24th Infantry Division is on alert and almost ready to go. The 1st Cavalry Division is just starting to get ready."

"What about the 25th and 7th divisions?"

"The 25th Division is in transition, meaning they have mostly recruits and are still getting organized. The 7th Division is a shell. I don't consider them a combat-ready unit. They're poorly led and only have a thousand men in each of their three regiments. We'll be ready to move with what we have as soon as we can, General."

"Oh," added Colonel Harris, "We just got in the first shipment of the brand new Super Bazookas. It's the only new Army weapon system developed since the war. As we learned in Europe, our old 2.5mm bazookas just bounced off German tanks. These fire 3.5mm rockets that will penetrate eleven inches of steel; they'll make short work of those T-34s. We only got fourteen of them in by air for training purposes. The rest of a limited supply are coming by ship."

"I'd like to get a few of those to our most important Korean defense Post," said Colonel Nelson. "If you add a 60mm mortar and plenty of ammo, I'll fly them into Seoul/Kimipo first thing in the morning, and yes, we'll keep it on the QT."

"That's a plan, Colonel! I'll get right on it."

"Gentlemen. That was good input and suggestions, now let's get ready to fight a war." General Walker hesitated, turned to Colonel Nelson. "May I have a word with you in my office?"

A few moments later, as they entered his office, he said, "I guess that Top Secret stuff is starting to pay off?"

"Oh, it sure is. We've got some wrinkles that we're working through, but yeah, it's keeping us well informed about anything happening near the border."

"Does anybody know about your relationship with the CIA and the use of their equipment?"

"Besides you, no. Colonel Harris now knows about the Posts and that we're using secret equipment, but he doesn't know any details or about the CIA."

"Make sure it stays that way. Some might think your reporting to the CIA as an act of disloyalty."

"I wouldn't dream of such a thing, General."

"Please, sit Jim."

General Walker walked over to his credenza by the wall, opened the bottom door, pulled out a bottle of Johnny Walker Red, and two glasses.

"They don't call me *Johnny* for nothing." He laughed, set the bottle and glasses on his desk, and sat in the chair next to Colonel Nelson. He leaned over and poured.

"I know you like yours neat."

"Is there any other way?" Nelson raised his glass to General Walker.

"To you, General. To your continued success!"

"Thanks, Jim. With your help, I hope to do just that."

They each took a good swig.

"Jim, I've known you almost six years now. You're as smart a man as I've ever met. Forget your degrees from Harvard; you have intuition, see the big picture, and yet can hone in on the smallest detail and realize its implications. I've seen what you can do and have benefited greatly from your council and your actions."

Nelson knew the general wanted something, reassurance maybe. He stayed quiet.

"Nobody in any position of power takes a North Korean invasion seriously. This military buildup could be just saber-rattling. Any significant action would have to be a political decision and approved by Stalin. Do you think he's crazy enough to give it a go?"

They both took a sip of scotch.

Leaning back in his seat, Nelson was reflective. "The people who expect this to happen have no voice. Politicians rule our State Department, and they have the President's ear. I've studied Russia and Stalin for the last ten years, and it's clear to me that Stalin is the most ruthless and cunning adversary America has ever faced. He makes Hitler look like a schoolboy bully."

"I agree with you about Stalin." The General took another sip.

Nelson continued.

"For the last one hundred or so years, Russia has wanted access to the Pacific to get a year-round ice-free port. They went to war with China in the late eighteen hundreds for this purpose. They didn't get it then, but they haven't stopped wanting it. They entered the war against the Japanese two weeks before it ended. Why? Because they wanted a seat at that table. Stalin was pissed that he didn't get any rights in Japan. I think we thought he would cause trouble in Europe, so we threw him a bone in Korea, and gave him half the country. Stalin was so happy that he probably pissed his pants. North Korea doesn't have any major ports, but South Korea does. Stalin sent over five thousand military advisers to North Korea right after the war to train their Army. They're still in the country. They sent hardware to arm this new Army. Granted, it was war surplus, but these weapons were some of the best in the War and still are."

The general nodded.

"Now comes the interesting part. Stalin's thinking is that America has the A-Bomb, but, so what? Will America use it to defend a nothing country against a civil war? Even if it's Russian-backed? I think he thinks we say no. Then, last January, in a speech to the National Press, our Secretary of State, Dean Acheson, outlined our strategic Asian Defense Perimeter, and he didn't mention South Korea. I don't think Stalin missed the implication. Whether an oversight or mistake, I believe that it reinforced Stalin's belief that South Korea is not of strategic importance to the United States, and that we wouldn't intervene if there were trouble."

"Maybe we won't come in?" The General refilled his glass. Nelson waved off his offer.

"You're right General, maybe the powers that be decide that Korea isn't worth more American lives. The public is war-weary and would raise a big stink. Politicians won't like that one bit."

Nelson paused for a moment, then continued.

"But I believe that, sometime soon, people in power are going to wake up and realize that if South Korea falls into the hands of these Commies, they'll realize that they'll only be eighty miles away from Japan. I think that'll scare the hell out of them."

Chapter Four

Joe walked into Rusty's office, shaking his head.

"I've been thinking about this guerrilla option. You called it special ops. On Okinawa, we called it survival."

"Right, that's what Colonel Nelson liked about you. Always volunteered to go out in the night and blow something up or do a snatch and grab. Well, it seems you're going to get that chance again."

"Looks that way. Gotta say, it sure beats kissing ass and polishing boots."

"This is going to be different than what you've experienced. Hell, it's going to be different from what we've all went through. Fighting a delaying action behind enemy lines most of the time. It's not going to be pretty, and we'll be at it for a while."

"So," Joe said, "if your plan is to survive the initial invasion, why did you build this Post as if it were the Alamo?"

Rusty grinned.

"Oh! Did I forget to mention our escape tunnel?"

"An escape tunnel!" Spreading his hands, Joe shouted, "Jesus! When the hell were you going to tell me! Where is it?!"

"Calm down. Too much going on. It's behind the southwest corner fortified emplacement, called forts. The tunnel is ten feet below the surface and extends out six hundred yards, two hundred yards beyond our initial barbed wire, ends in a large scrub brush area where there's a hidden trap door. The tunnel is a series of thirty-six-inch wide concrete water main pipes. We use it for nighttime reconnaissance outside the wire, and, of course, it's our way out when we have outlived our stay here."

"Son-of-a-bitch! You have no idea how much better I feel. This might sound stupid, but how in the hell did you build it in this ungodly place?"

"It's not a stupid question, and it's a good story. Colonel Nelson, Red, and I planned this particular Post because it was going to be the most crucial after the initial attack. First off, it needed to survive a surprise attack. Then, if an all-out invasion started, it needed to hold off a determined attack for at least a day, hopefully, more. But, of course, we then had to have a way to escape.

"This brings us back to the beginning. The Army had no budget for this installation. Besides, they wouldn't have approved it anyway. This is a Top Secret operation. Anyone who knows we're here thinks we're a special training unit, working with the ROK Army. Our colonel is highly resourceful, has a lot of pull with a lot of people. So he rigged contracts for the concrete and steel from Seoul contractors. He also got the Korean President, Mr. Ree, to pad the contract for a new water pipe being constructed on the northern side of Seoul so that we could get this pipe. Of course, the US of A was paying for all of this anyway. Nelson wanted to keep everything about this Post secret, particularly the pipe, so he had the pipe shipped nightly. After each delivery, we placed the tube with heavy equipment the ROK construction platoon left with us. The trucks left before morning light. It took us ten days to finish placing those monster pipes."

"Impressive! A lot of good planning went into this Post. This colonel sounds pretty smart."

"Yeah, Joe. He is." Rusty's stomach growled.

"Come on, let's get some grub. I'm starving." Rusty got up. "You'll meet the rest of the team; they're off security duty now."

They walked down the hallway and entered the kitchen/mess hall. A large pot was cooking on a corner gas field stove, steam rising. A pleasant aroma filled the room. A photograph on the floor was playing a Bing Crosby song, *Faraway Places.* Four men were sitting at a makeshift table, talking and eating from bowls.

"Ain't this cozy."

"Guys, say hi to Joe. Joe, this is Sergeant… hell! They're all sergeants! That's Arron Knight," pointing to the tall skinny guy on the right, "next to him is Javier Salinise. Marty Fells next to him, and you've met Tracker."

"If you gentlemen don't mind, we'll join you."

"Oh crap, now the neighborhood is really going to shit." Arron quipped.

"Pay no attention to Arron, Joe, he's from the Everglades or some other swamp in Florida and wouldn't know a neighborhood from an igloo," Rusty said, smiling.

"Now you just hold on there, Mr. Yankee carpetbagger. Remember, I made that stew you're about to eat."

Rusty and Joe sat down at the table.

"This stew is delicious! If you cook like this every day, I may actually gain some weight."

"Damn it, Joe, don't give Arron a bigger head than he already has. He's hard enough to deal with as is." Rusty shook his head.

"So Joe, I hear you're a Yankee, too. Where from?"

Joe couldn't help himself. "Started sharecropping in Flatbush, Brooklyn." The men laughed.

"Oh man, another Yankee comedian," Arron went on, "but at least we have something in common."

"What's that?"

"I was raised on a farm just outside of Miami. My family did a lot of sharecropping."

Joe laughed.

"The only sharing of crops I did was stealing vegetables off of the street vendors on Flatbush Ave."

Marty spoke up, "I'm from Bridgeport, Connecticut, met a lot of Brooklyn guys that spoke real funny. You don't, how come?"

"Thanks for the compliment. It's taken me a long time to learn how to speak American. I left Brooklyn for the first time when I went off to boot camp in California. Realized I spoke funny, people kidded me about it. I didn't like that; I just wanted to fit in, not be laughed at. So I started mimicking how other guys spoke, especially guys from the Midwest. I think they speak American the best."

"Wait a minute!" Tracker scratched his head, "I'm the only real American here! After stealing my land, you now insult me by talking to me in a foreign language and claiming it's American. Don't that beat all?!" Everyone laughed.

"OK. Time to get serious." Rusty got up from the table. "Tracker, what's your plan for tonight?"

"Sorry, Rusty, I was going to talk to you earlier but got sidetracked. It's simple, and Red agrees with my idea. I'll take Foxy and Snake. Oh, Joe, that's Javier's moniker because he's so sneaky, ghost-like."

Javier smiled.

"We'll go out through the tunnel after dark and wait near the outside northwestern wire. I think that's where they'll be next. We'll jump 'em Indian style, maybe scalp one or two, bring in any survivors for a serious discussion."

"You sure have a way with words, Tracker." Rusty was smiling even though he knew how dangerous this mission was. "No chances, right? If you can't take any alive, that's okay. We'll get info off their clothes and papers. No hesitation. You read me?"

"Yeah, dad." Tracker and Snake said at the same time. "We'll leave at 2200. Foxy will finish his security detail in about an hour; we'll have time to get set up."

"Hey, Tracker! What about me?" Marty said. "You could use a fourth guy in case things get bad."

"Don't worry, Marty. You'll get your chance before our little tribe makes peace with these Commie folks. Besides, the only thing that's going to be bad tonight is going to be what happens to our uninvited guests."

Joe broke the momentary silence.

35

"Hey, Marty, Red mentioned you're our medic. How long you've been at it?"

"I've been doing this going on six years. My first and last action was on Okinawa."

"No, shit!" Joe said, amazed. "I was in the 96th Division, what outfit were you in?"

"Damn! I fought right next to you in the 7th Division."

The two men stared at each other. Neither said a word.

Sensing the need to say something, Rusty chimed in.

"Marty has some balls, Joe. He got a Bronze Star for defending an aid station that was about to be overrun." Rusty felt awkward about promoting Marty; he didn't need to.

Both Marty and Joe ignored Rusty's interruption. The room went silent again, so Rusty changed the subject. He turned to Joe.

"Did Red tell you you're on radio duty tonight?"

"Yeah, I got my schedule. I'll be on special alert tonight."

"We'll all be." Rusty sighed.

Korea - The Post - At the Tunnel – 2200 – (10:00 pm)

Behind corner fort # 3, the three-man team gathered by the tunnel entrance. Red was pulling security duty at the emplacement. Foxy pulled back the ground camouflage cover over the trap door and pushed back the sliding rod to release it. He held the thin rope attached to the door, so it wouldn't slam down and make noise, and gradually let the line out until it was open.

"Snake, go first, then you, Foxy," Tracker said.

Each carried a Tommy gun, a .45 caliber 1911 automatic pistol in a chest holster, along with an assortment of knives and a flashlight.

They climbed down the ten-foot ladder, entered the tunnel on hands and knees. Learning from the first time they crawled on their knees through the concrete pipe, their knees had protective wrapping of cloth padding for the six hundred yard crawl. They traveled light, no web belt, no helmet, or anything extra that might make noise. They moved quickly through the tunnel.

The exit shaft was four by four wide and ten feet high, the same as the entrance. Only two men could stand at the base of the ladder. Tracker waited in the tunnel as Snake and Foxy went about opening the trap door. Snake climbed the ladder. With one hand, he slid back the rod, holding the door. With the other, he held it so it wouldn't slam down on his head. He backed down a few feet, let the door fully open. Then went back up to the camouflage covering on top of the thin steel sheet, unhooked it, and lifted it high enough to slide it to the left, just enough to clear the opening.

He climbed to the top, looked around, got still, and listened. Nothing. The heavy cloud cover made the night almost black. Perfect, he thought.

He whispered down the shaft, "All clear."

Emerging, they paused, got their bearings, then headed to the corner as planned. This was the tricky part. A jagged edge of the mountain jutted out to the first wire, making access impossible. Anticipating a clandestine approach to the north from this rear corner, Red had carved out handholds into the cliff so the men could climb over the barbed wire. Barely detectable by day, impossible to see by night, it was the perfect solution for what they were planning tonight.

Up the side of the cliff and over the barbed wire they went, then moved along the tree line just outside and along the barbed wire to the northwestern corner.

"OK, I think this is a good spot," Tracker whispered, "They should come from the left. Just to the edge of the tree line. That's where we'll take them as they're getting into position."

They fanned out left and right, each about four feet apart, then settled in behind trees. They drew .45s and flashlights., and got perfectly still.

Two hours had passed before they heard movement. Three men hunched low, walked along the tree line, just as Tracker predicted.

As planned, Foxy on the right would take the point man. On the left, Snake would take the third man. They figured they'd be too far away to jump them without making noise, so the plan was to shine flashlights on them while Foxy yelled in Korean, "don't move or you die!"

Tracker signaled 'go,' stepped away from his tree, and shined his flashlight on the middle man. Foxy and Snake did the same as Foxy yelled his warning. Foxy's man turned his rifle towards him; Foxy fired. The other two froze and dropped their weapons. Foxy shouted, "Hands up!"

They searched them, removed weapons, tied their hands behind their back. Foxy could see they were terrified. They looked like boys. Difficult to tell, he thought. Koreans all looked young. He sighed, such waste.

"I'll drag the body into the trees. Stash the weapons. You guys head back," Snake said, "I'll take care of closing the tunnel."

They set off with their prisoners, reentering the post the long way via the front gate so the tunnel wouldn't be compromised in case the two escaped.

Korea - The Post - Thursday – June 22, 1950
Rusty's Office – 0030 - (12:30 am)

"Any Intel from the prisoners?" Rusty asked.

Foxy ran his hands through his hair, sat back in his chair.

"Poor bastards, they've been through the mill. The one I killed was a political officer. A pretty stupid one, as it turned out. He made sure these guys did their job. They were thinking about killing him and deserting. I believe them."

"When did they get here?"

"Their regiment arrived two days ago. They are the first element of the 6th Division. They'll be the lead division if there is an attack. They said that this is the North's most experienced division, largely made up of veterans from the Chink's war against the Japs.

They fought four more years fighting the Chinese Nationalists. After they kicked the Nationalists out, the Chinese started to disband a good chunk of their Army. They rounded up all their Korean volunteers and conscripts. Sent them back to North Korea, who were starting to build up their own Army."

"So? Who are these guys?"

"These two escaped to China in 1944 when the Japs were taking all able-bodied Korean men for slave labor. They were more than willing to fight the Japs for the Chinese. But after that, they had no choice. Both are from small villages up near the Yalu River. They figured they could make it back there if given a chance. They figured this was their chance."

"Any tactical info?"

"Nope. All they were told was that this is a big exercise in case the South Koreans should invade them. Their mission was to find out what forces were deployed near the border. They reconnoitered three miles east, west, and south of here. They were heading back once they had finished with this Post. Their political officer was very curious about this facility and wanted to find out more about it. The good news is that they have had no communication with their base unit since they left."

"Did they see any heavy stuff when they came in?"

"They saw artillery."

"Good job, Foxy. Get some sleep. I need you to go to Kimipo in a few hours to bring Hank's body to Billy. Also, transport is arriving at 0900 with new weapons. I'll need you to pick them up. Take Joe. On the way, deliver our prisoners. Drop them off at the ROK 1st Division headquarters."

"Come on Rusty, if I drop these guys off there, you know what will happen."

"That's not our responsibility, Foxy. What else can we do with them?"

"I could drop them off at South Korean Army headquarters in Seoul. It's only a little out of the way. I've met Colonel Pyong Oh, their G2 Intelligence boss. He's a reasonable guy. I could talk to him. Maybe he'll spare their lives. Maybe they'll have a chance."

"You've got a good heart, Foxy. You're on. Don't have a lot of time, though. Colonel Nelson says things are moving quickly. I want the new weapons he's sent us ASAP."

"It won't take long. Do we know what he's sending?"

"No, the cable only said weapons and ammo."

Later, Joe got off duty, was tired, and went to the sleeping quarters. Most of the other guys were still pretty hopped-up by the night's success and were in the mess area. Marty was alone in the room, lying in his bunk.

Joe called over, "The guys did a great job tonight."

"They did, thank God. I was hoping I wouldn't have to stitch anybody up or worse."

"Marty, if you don't mind me asking, why did you stay in after that horror show?"

"Probably the same reason you did. First off, I love what I do. Then I realized I had a gift. I saved a lot of men on Okinawa. I felt important for the first time in my life. Knowing I'm good at something gave me a feeling I'd never experienced in my life, and didn't want to give it up. For some reason, I have an exceptional ability to treat the wounded and" pausing "I find redemption helping men die peacefully." He grunted softly, "If that's even possible."

Joe propped himself up.

"You know, just talking to you about that place brings up a lot of memories that I've tried hard to forget. You're right. That place changed my life. Everything before became meaningless." Joe paused, his mind flashed on a recurring memory, two men that he failed to save that day. He shook his head. "Anyway, after that, I decided that I 'd never go back to Flatbush Ave."

Marty knew he should change the subject.

"So, what did you think of Colonel Nelson?"

Joe laughed, proceeded to tell his story. Marty listened, smiled, and said, "That Colonel is something. Too bad you didn't get to meet him."

"So, how'd you get here, Marty? It sounds like you volunteered."

"Let's just say our colonel is very persuasive. After the war, when my unit got stationed near Nagasaki, I was thinking about getting out and maybe trying to get into medical school. But then I got involved with the local Japanese. The Atomic Bomb had dropped fifteen miles from our base, so you didn't have to go far to see the effects. The locals took in large numbers of wounded, tried to make due. They were overwhelmed. I decided to help. Each day I went out there and did what I could. My CO found out what I was doing and wanted to help out, too. While he hated Jap soldiers, he didn't think civilians should suffer. It wasn't long before we had a field hospital set up in town, and our Regimental Surgeon, Colonel Holly, was showing up every damn day. He was a great guy. I learned so much from him."

Joe realized Marty was much more, way more than he initially thought.

"So why didn't you get out and go to medical school?"

"Eight months ago, I put in a request to leave. I figured I'd done enough. A few days later, Colonel Nelson came to me. He told me about this team he was putting together, its mission, and how important I would be to its success. If I agreed to join him for one year, he'd immediately promote me to sergeant, and after the year was up, he'd do everything he could to get me into Harvard Medical School. As an Alumnus, he's got some pull."

Joe was impressed. "That's quite a story, Marty. Glad you're here."

"Thanks, Joe. Glad you're here, too. A miracle we both survived that damn meat grinder." Hoping for a few hours of sleep, they turned in.

Chapter Five

Foxy and Joe finished hooking up the trailer, laid a blanket over Hank's body, secured him to the trailers side rail so he wouldn't get thrown out on the bumpy ride. Their prisoners were seated in the back of the jeep, hands, and feet tied.

"You ready?"

"Yep, always ready," Joe said as he swung his Tommy gun into a firing position across his lap. They headed out.

"This road's in pretty good shape now. The spring was dry, and the dirt's compacted real tight, so it doesn't have many potholes. That'll change in a few days when the summer rains start. They call it monsoon season. We got here last year at the tail end of it, and it was rough. This road gets horrible. Down about a mile or so is a small bridge over a dry stream. When the rains start, that stream becomes a river, a natural defensive position on the south side. I saw a South Korean company set up there yesterday. We'll stop on the way back and check them out."

A while later, they came upon a mangled burnt-out truck off to the side of the road.

Foxy slowed down. "This it?"

"Yeah."

"You were lucky, my friend."

Two hours later, just outside of ROK Army headquarters west of Seoul.

"How'd it go?" Joe asked as Foxy got in the jeep.

"Colonel Oh was in a bad mood but appreciated that we didn't kill these guys. He happens to have a friend that lived near the town where one prisoner grew up. They talked about the area. He believes their story, was happy to get some new intelligence. He thinks he can keep them alive. Colonel Oh was upset about losing a three-man team he sent up to cross the border last night. They were spotted before they got very far. Their ROK border unit saw the whole thing. At least the colonel learned that the NK line in that sector had just gotten heavily reinforced."

"Did he say anything more about what's happening?"

Foxy had just exited the gate and turned toward the airport.

"Just that we should be very alert. ROK units have run into a lot of infiltrators along the road north to the border. I told him about your meeting yesterday."

On the airbase, they approached the Army C-47 transport.

"Look for a bald fat guy, a sergeant."

"Yeah, I know, met him yesterday. Charming fellow."

"There he is." Foxy pointed to the driver of a forklift to their right. He waved at him. Billy didn't wave back.

They drove up alongside the plane. Billy pulled up next to them and got off the forklift and went to the jeep's trailer and lifted the blanket.

"Poor bastard." He said to Foxy, "I've got a body bag, and if you'd help me, we'll get him in before he starts to smell. I've already cleared a spot on the plane for him."

They went to work.

"Your crates are next Foxy," Billy growled, acting as if nothing happened.

"Yeah, and I love you too. Just put'em next to the trailer."

"Sorry, Foxy. I'm pissed off. Shorthanded now, and I'm in a hurry to get this bird unloaded. Got a load to get over to the South

43

Korean 2nd Division headquarters, twenty miles from Seoul, and it's a pain in the ass road. Gotta be back before dark."

"Everybody's on edge, Billy." Foxy couldn't believe this guy. "Let's just get her unloaded. We're in a hurry, too."

Billy placed four large crates, each one was four feet by four feet by eight feet long, along with ten smaller boxes, next to the jeep. As he finished placing these, he said, "Big damn load for what you got to carry it in. Oh yeah, there's a loose M-60 mortar for you, too, and more ammo crates. Be right back."

Foxy said, "Okay, let's get these crates opened."

Joe grabbed a trenching shovel off the side of the jeep, started on the big box.

"Man! Look at this bazooka! Never seen anything like this baby!" Joe took out two tubes, fiddled with them, and connected them. "Damn! Must be five feet long."

Foxy nodded. "Impressive. We had those crappy 2.5mm bazookas; they could hardly stop a truck, never mind tanks."

Joe shook his head. "We used them to knock out Jap bunkers; they were shit. Luckily we didn't see many tanks."

They stowed the weapons in the trailer, took all the ammo out of the crates and placed them wherever they would fit, filling the trailer and the rear of the jeep. The mortar was last and went on top of the pile in the trailer, and then they tied everything down.

"That went better than I expected it would." Foxy grinned. "We'll take a small detour, visit my *Halmoni*."

"Huh?" Joe asked.

"That's grandma in Korean."

"Does that mean we can get something to eat?" Joe smiled and looked hopeful.

"That's why we're going!" Foxy laughed.

An hour later, as they drove out of Seoul, Joe said, "I don't know what the hell I just ate, but it was great! I wish I could understand what you two were talking about. You were having a ball. She never stopped smiling! I like your grandma."

"She has always been a happy person. My mother tells wonderful stories about her. I'm so glad to have met her, really love seeing her often. Her cooking is great, right. I told you you'd like it. Too bad that my cousin was out getting supplies. You'll meet him next time."

The road had light traffic. Dry rice paddies on each side stretched as far as one could see, towering mountain ranges loomed in the distance. They passed the occasional shack on the edge of the road. Further ahead, the road veered closer to the hills. The surroundings changed to high grass and tall bushes, a familiar setting.

"This is a beautiful country if you don't look at the poverty," Joe commented. "But I guess you could say that about a lot of places. Coming from Brooklyn, though, and seeing open spaces like this ….well, it's breathtaking."

"Chicago had everything I wanted when I grew up. We lived just outside the city, beautiful rolling hills, close to Lake Michigan. I didn't see poverty until I started going into the city with Dad. I thought it was a city thing. He explained it to me, but I really didn't get it until I was older. Even prejudice eluded me. Being a Korean with a white physique made me popular with my school mates, unique in a strange way, I guess. The Army educated me on the realities of life, however. That was an eye-opener."

Joe looked at Foxy as if he were from Mars. "Beautiful rolling hills? Near Lake Michigan? Can't even imagine it."

Ten minutes past Joe's ambush site, they rounded a bend in the road, going slow on the slight uphill pitch because of the load they carried. A farmer lay in the road next to his bloodied dead Oxen.

Foxy was angry. "Looks like those bastards picked on a local this time."

They stopped about twenty feet away. Foxy got out, started to walk towards them when Joe fired his Tommy gun in a quick three-second burst. Foxy dove to the dirt road and looked up in time to see the prone farmer start to rise with an automatic rifle in his grip.

Then the farmer's head exploded as bullets from Joe's second three-second burst impacted.

"Holy shit!" Foxy froze. "Are we clear?"

Joe got out of the jeep, walked into the high grass off the road. "I think so, but I'm going to make sure. Stay down." He went over to where he'd fired the first volley and found two bodies, their weapons on the ground. He semi-circled and found another body with no weapon.

"It's clear," he yelled, "two dead bad guys and a dead farmer."

Foxy walked over to Joe as he searched the bodies.

"How'd you know? Did you see'em?"

"I didn't. When I get this tingling feeling, I've learned that something dangerous is around. When we stopped, I got that feeling. I thought that **if** I were going to stage an ambush, I would position myself in this high grass right here. So, what the hell, I thought, just open fire; what's the worst that could happen? Scare the crap out of you?"

"You did….and saved my life." Foxy was still a little shaken. "Thanks, Joe! Let's get out of here."

"I'm not sure if you're a good luck charm or some magnet for trouble."

"Think charm!"

Pulling out, they passed the dead oxen. Foxy laughed.

"What's funny?"

"Too bad Arron isn't here."

"Why?"

"You'll learn Arron's a genius with roadkill. We could be eating oxen steaks for a week."

They decided not to stop at the ROK position by the stream. When they arrived, Arron was on duty at the Post gate entrance.

"Nice seeing you two. How were the girls in Seoul?" He said with a smirk.

"You're an asshole, Arron! You just can't help yourself, can you?" Foxy yelled back.

"Jesus, what the hell did I say?" Arron looked like a schoolboy caught doing something wrong.

As they parked, Red appeared.

"Glad you're back safe and brought us some goodies too." He eyed the weapons and ammo. "Any problems?"

"Nothing Joe couldn't handle."

Foxy told Red what happened.

"Jesus! You got some good instincts, Joe. Good job! Glad you saved this sorry slanted eye character, I've become quite fond of him."

"My Brooklyn upbringing, Red. It's finally paying off in spades."

Red smiled. "Oh shit, there are way too many New Yorkers around here." The men chuckled.

"Let's get these weapons cleaned and set up, and oh, Foxy, find me the manual on the bazooka. We'll have a class on it later. After it's cleaned, stow the mortar in the storage room. We'll deal with it tomorrow."

Foxy wanted to know. "Red, do we need four bazookas?"

"Definitely two. I'll ask Rusty what he wants to do."

Red sat down in Rusty's office.

"The boys had a situation on the way back." Red filled Rusty in.

"Thank God they were alert." Rusty was reflective. "Joe's file had a few references to his uncanny ability to sense danger. He saved more than a few men on Okinawa. Sure happy to have him."

"Affirmative. They brought in four of the new Super Bazookas we heard about plus a total of forty-two 3.5mm rockets. Do you want to keep all four weapons? I thought we'd keep two and give two to the ROK troops nearby. What do you say?"

"Good idea. Train all the men today. Tomorrow we'll decide who'll get them. I was thinking of sending Foxy up to see Captain Hui at the border. Also, I want him to check out that new South Korean unit by the stream. They might be a better choice for these weapons."

"Good plan. Let's see what's going on later and decide. Who'll be riding shotgun with Foxy?"

"You call it."

Same Day – Thursday, June 22, 1950
Two hours later, near the northwest corner of sandbagged fort # 1.

Red was teaching the second group of four men on the basics of the new Super Bazooka.

"OK, let's wrap it up. You've learned how to set it up, load and arm it, aim it, and how to fire it. I've told you about the recoil and the backblast danger of the rocket. I repeat what I said earlier about using it. Whether a tank or truck, it will have supporting fire. So NEVER, EVER, expose yourself. You may have seen your buddies use it in battle or training somewhere, setting up in a kneeling position to get a better shot. That will get you killed, as it did a lot of guys using the textbook method. Forget it. Some engineer was finally smart enough to include a bipod for this unit so you can lay prone to fire it. That's the only way you fire this weapon in the field." Pointing to the fort behind them, he said, "If you're behind a structure like that, with a protected slit opening, fine. Otherwise, lay prone. Clear!?"

The men nodded, Arron said, "Yes, mother."

Red smiled. "This weapon is a tank stopper. It will destroy anything that moves if you get a good shot off. You'll have time to do that because it has a thousand-yard range. At five hundred yards or closer, it'll give you your best shot, so don't rush it. Any questions?"

Arron raised his hand. "Tanks scare the crap out of me, Red. Just hearing them makes me want to piss my pants! I'm not going to wait too long for that son of a bitch. If it gets within range at nine hundred and fifty yards, I'm firing." The guys chuckled.

"Glad you told me that Arron. You won't be assigned to first duty on the new weapon. But let me tell you something, you still may have to fire it at some point, though. I want you to think about missing that tank from that range, and then you have no rockets left."

On that note, the men dispersed.

At his radar console, Ray yelled, "I'm picking up two bogeys! Fifty miles out! Heading due south towards us!"

Rusty yelled, "Sound the air raid alarm!"

Ray hit the switch, and the siren blared. The men scrambled, knowing exactly how to respond. Grabbing helmets and boxes of ammunition, ran outside, and took their positions at the forts.

"What's their ETA, Ray?"

"At present speed, about eight minutes. Must be props, they're going about two-fifty."

Rusty said, "Joe! Get a quick flash message off to Colonel Nelson! Tell him what's happening. It's a wait and see. They haven't crossed the border yet, so we don't know what their intent is."

Each fort had a thirty-caliber machine gun, mounted on a stand with a swivel base. Positions were numbered; on the Northside, from west to east, were numbers one and two. On the Southside, position three was on the Southwest corner, and four was just outside the command post entrance.

Communications to each fort came from a hard-wired two-way system tied to the inside of the command post. Rusty grabbed the phone and hit the switch for all forts to hear."Listen up! Enemy planes are approaching from the north! Commence firing if they come into range. ETA five minutes."

The four crews loaded tracer rounds; clicking sounds filled the air as the men primed their machine guns.

"They're at the border now! They're crossing in! Holy shit!" Ray shouted.

"Joe! Tell Nelson, these fuckers crossed the border!"

Rusty stood behind Ray and watched the radar screen, mike in hand. "Forts one and two! Two planes are coming your way! They're flying straight at you at five hundred feet! Speed, two-fifty,

ETA less than 3 minutes. Forts three and four! They're going to pass right over you: so start firing up their ass when you hear forts one and two open up! Let's get'em!"

Each position had a two-man team except fort number four, just outside the south entrance. It was one man short, Rusty.

The machine guns were belt-fed. Firing at five hundred rounds per minute, it needed a feeder to keep the ammo from jamming.

Rusty had practiced running to his position, got it down to twenty seconds.

"Joe! You're in charge! Ray! Keep your eyes on the scope! I want to know what they do after they pass by. I'm going out to four."

Manning number one position with Arron, Red heard the plane's engines. Then dots appeared low in the sky. The order had been given to fire when the opportunity was right. Red was steadying the ammo belt for Aaron.

"Go, Arron!"

Fire erupted. The planes were just out of range, but shooting a little early allowed for better anticipation of aim. With a fifteen hundred yard range, the thirty caliber machine gun was an exceptional weapon. At a thousand yards, the planes could be easily hit if it was anticipated and tracked correctly.

Foxy and Tracker at fort two opened up right after number one. They had a good bead on the second plane to the right. Foxy watched the tracer rounds go up from number one to the plane on the left. Each position was aiming at a separate aircraft. Good, he thought. Tracker was shooting, and he was feeding. They would only have time to fire one belt, so he didn't have to worry about getting the next belt ready. They would have less than thirty seconds of clear firing until the planes passed.

The enemy flew directly overhead. Number three and number four guns opened up at the same time as the first guns. Smoke started coming out of the aircraft that flew over number four position. Then they were out of sight.

Marty yelled, "I got one!"

Rusty grinned at Marty. "I know you think being Jewish and all makes you unique, but you didn't even come close to hitting that plane!"

"God damn it, Rusty! I am special!" Marty jabbed him in the shoulder. Rusty laughed.

Joe came on the mike. "These guys are starting to veer west. The southernmost plane is losing altitude. He's down to two hundred feet. The other plane is making a slow turn. Hold on."

Within seconds, Joe came back on. "I lost sight of the southern plane. The other plane is heading due north, maintaining altitude at five hundred feet. It looks like he's going back to the barn. We got one. Good shooting!"

Celebrating, they slapped each other on their helmets.

As he walked into the command post, Rusty asked Ray, "Where are they now?"

"The one we lost sight of I'm sure is down. The other has re-crossed the border, heading north. I'm just about to lose him now."

"Any idea where that plane went down?"

"Got an idea. Guessing five miles southwest of here, but I don't know how long he glided. He could be anywhere in a ten-mile radius."

"Joe, request Colonel Nelson alert a South Korean unit to do a recon of that area. Find out if that plane had surveillance cameras."

Rusty grabbed the mike. "Guys! Great shooting! Stand down! Threat neutralized! Clean those weapons and come on in."

Cleaning their machine guns, Arron paused, looked at Red, "You being from north of the Mason-Dickson line and all, makes it hard to say, but what you did to build these forts and jerry-rig these machine guns, well... it's just plain genius."

Red chuckled. "I don't think I ever heard you compliment a northerner. I wish I could take credit, but I owe this inspiration to the krauts. They made these forts and modified weapons like rabbits have babies. That's where I got the idea to take the machine gun mounts out of the jeeps. It's the same thing with these steel protective plates on the sides of these weapons I welded on. It

51

pissed me off that those side-mounted plates they used kept us from killing the gunner. I figured we could do the same thing now. Protect our guys in the same way."

"No matter. You figured it out and did it. My statement stands." Arron went back to work.

Just as it got dark, Red came into Rusty's office.

"What do you think, boss? They didn't shoot at us; they didn't bomb us. Was it a recon mission?"

"That's my bet. They came down the expected main line of attack. I suspect they were trying to figure out if the ROK have any aircraft to meet them or if they had any heavy weapons near. One thing's for sure if they didn't know about us before, they sure as hell do now."

Chapter Six

Foxy waited for Joe in the jeep when Rusty walked up.

"Don't go off the beaten path, Foxy. Just check out the border, see what's up with Captain Hui. Come back and report before you go down to the ROK position by the stream."

"Got it. Thanks for letting Joe come along."

"That was Red's call. Now, don't you get all superstitious about Joe being a good luck charm! Everyone here is more than capable."

"Okay. I'll think about it."

On his way out of the building, Joe passed Ray and noticed a group of wired switches on the side of his desk that he'd not seen before. Looking closer, he saw that there were three rows of ten switches each, all numbered, one to thirty. It was quite a setup.

He pointed at it. "Hey, Ray, what the hell is this?"

Ray grinned. I should have shone this to you yesterday, but you were busy. He then pulled out a hand-drawn map.

"This is a little surprise for any unwanted visitors. See the road?" He pointed at the illustration. It showed the road in front of their compound dotted with numbers, starting with #1 on the far northern end and ascending to #20 just in front of the gate entrance, then proceeding up the road toward the compound, ending in #30 about fifty yards from our door.

"We planted two anti-tank mines per number, Joe. Disabling the pressure fuse, me and Larry wired them up with C-4, dug the wires to this little electric switch gizmo. Pretty cool, eh?"

Walking out of the compound, Joe smiled, threw two canteens in the back of the jeep.

"Why're you smiling?"

"Ray showed me his toy switches."

"Yeah, Santa's deadly little toy shop. He and Larry do good work!"

Japan – Eighth Army Headquarters – Colonel Nelson's Office - Same Time - 0700 - (7:00 am)

Nelson was reviewing reports when his phone rang. It was his duty sergeant at the front desk.

"Good morning, Colonel, I have a first sergeant Army Ranger named Dan Tully here requesting permission to speak with you."

"Did he say what it's about?"

"Yes, sir! It concerns a mission."

"Okay. I'll see him."

The Ranger walked in, stopped, and snapped a salute.

"Colonel, thank you for seeing me."

"First Sergeant," the colonel saw his nametag, "Tully. What's this mission you want to talk about?"

"Sir, may I speak freely?"

"Permission to speak. At ease."

"Thank you, sir. I know this is unusual. I may be out of line…"

"Don't worry; I'll let you know if you are."

"My Ranger company is attached to the 25th Division. We have all new officers with no practical knowledge. My squad is made up of veterans with combat experience, and all desire a change. We don't fit in, not within our company, and certainly not with the 25th Division. We've collectively gotten into a lot of trouble, raising hell and such. Our regular Army officers have us, and our company, on

54

the shit list. We were alerted yesterday to standby for potential deployment. The squad got to talking about our situation. We decided that our role in any future action would probably be minimal at best. Should it be more, though, we worried about our leadership. What we've seen so far makes us feel that we wouldn't last too long in a real fight."

"Son, you're walking on some damn dangerous ground here. Get to the point. And quickly!"

"Sir, with all due respect! My squad is exceptional; we're combat ready and seasoned. Right now, we're nowhere and going nowhere fast."

"Go on."

"Sir, I know you've got a special mission going. Things are starting to happen. Maybe you could use my squad?"

The colonel looked at him oddly, "Interesting. What special mission?"

"Sir, I met First Sergeant Red Kenny at a weapons analysis review study last year. They had veterans from all different units evaluating the various infantry weapons used by all the countries in the War. Anyway, we became friends. He told me about you, and his new assignment in Korea before he shipped out. My squad and I figured that maybe you could use a little more help. At the same time, this would get us the hell out of Japan."

"Fascinating." Colonel Nelson paused and stared at Tully. "Your entire squad has agreed to this?"

"Yes, sir. Every man."

"Let me get this straight. You flagrantly chose to usurp the Army's command structure protocol, disparage your superior officers, and put in for an assignment that you know nothing about."

Tully looked straight into Nelson's eyes. "Yes, sir. That's exactly right."

"You've got balls, Tully; I'll give you that."

Nelson paused to reflect. "I'm often reminded of how small the world is. You just confirmed that belief again. Another phrase comes to mind. "Things happen for a reason.""

55

Tully was puzzled.

Nelson went on. "Circumstances have become very fluid in Korea. I was thinking about making some different moves. You might be a piece of that change."

Tully brightened. "That would be great, sir!"

"I need to think about this, do some checking. I do have to go through channels, so no promises. But, if this works out, you could be on a plane in the morning. Your squad ready?"

"Colonel, we can be ready in twenty minutes."

"I'll let you know later today." Colonel Nelson stood.

Tully saluted. "Thank you, Colonel." He looked unsure.

Nelson understood. "Don't worry, son; I won't say a word about this conversation to anyone." And he added, "I expect you and your men to do the same."

"Affirmative, sir."

Later, Nelson drove over to the 25th Infantry Division headquarters to see General "Bill" Kean, its commander.

"Good morning, General."

Their friendship was forged in a history that had sealed mutual respect.

"Hi, Jim. Good to see you. Come in, sit. Coffee?"

"No, thanks. Had my three cups already, just wanted to stop by, see how your alert preparations are going?"

"I told Colonel Harris earlier that we're ready to go with what little we have and wondered if he knew how we might get there. I also asked where we might land? What our objectives might be? He didn't have much to say about these concerns, except that we should plan for the worst-case scenario. I asked him what that might be? He laughed, said I should think about the swift collapse of the South Korean army and be ready to land anywhere south of Seoul."

"I know about all the issues, General."

"I know you know. Probably more than anyone. I'm just blowing off steam. I've got my guys working on it. We'll be ready for any assignment. I wish we could get more troops. My division is more like a regiment. It's a damn shame!"

"We're all shorthanded, frustrated, but I'm glad you're at least planning. You'll probably not get involved for a while until your division can get up to strength."

Disheartened, the general nodded. "We're a proud unit. What's happened to this division is criminal." He paused, then said, "I guess you could say that about the entire Eighth Army!"

"You could." The colonel nodded. "We deal with what we have, General."

"Do you think they'll invade?"

"I do, and I believe that it'll happen soon. We don't know what Washington will do, or even if they'll do anything."

"You just focus on getting me good intelligence."

"Yes, sir."

The colonel was about to get up, hesitated, and sat back down.

"It may be premature, but I want to find out about a unit that's attached to your command."

The general squinted, gave a wry smile, "You want something, Jim. What is it?"

"You have a Ranger company attached to your command. I'm interested in a squad within this unit if you could fill me in on it."

"Sure. It's the reconstituted 2nd Ranger Battalion from the War. Your squad is from the 1st Ranger Company. The Army Chief of Staff wants to build the Rangers into companies that are attached to Infantry units. This unit is the first company to be formed. The Army hasn't pushed this yet, and I've had no further directives. Nothing has been done with this unit. Unfortunately, its leadership is lacking, as well. Why are you interested?"

"There's a squad in that company that I'm considering for a particular mission. Would you be okay with that?"

"What? Ten men? As I said, that unit needs to be upgraded. Until that's accomplished, they're of no use to me. If you want a squad, let me know."

"Thanks, General. If I could, I'd like to review the personnel files for this squad before I make a decision."

The general called out, "Sergeant Heinz!"

The sergeant stuck his head in the door. "Sir?"

"Colonel Nelson needs to see our personnel files for the 1st Ranger Company."

Colonel Nelson stood. "Thanks, and good luck, Bill."

He followed the sergeant to the file room, where he found the cabinet where the Ranger unit files were stored.

He searched for First Sergeant Tully's squad files, found all of the men. Then he went to a table, sat down, and opened Tully's file first.

Joined in '42, after extensive training and exceptional recommendations, volunteered for, and was accepted for Army Ranger duty. Was assigned to 2nd Ranger Battalion, action in Normandy, scaling the cliffs of Pointe Du Hoc, winning the Bronze Star for his courage. Additional action in the Huertgen Forest. Led his unit that captured the Castle Hill in Bergstein, where he was awarded the Silver Star. His last duty ended when his unit was relieved at Bastogne, where he earned a Purple Heart for his injuries.

Nelson closed the file. He knew all of these battles, how fierce they were. He was impressed with Tully, and thought, I want this guy. I'm guessing the rest of his squad will impress as well. He read each of their files anyway, and they sure did.

Chapter Seven

"How far we going?" Joe asked, just as Foxy turned north after clearing the gate.

"Three miles to the border, but we're going to visit Captain Hui. His headquarters is another mile to the west."

As they got closer, the terrain became rough, rock outcroppings everywhere amid lush mature bushes and trees.

They were flagged down by a South Korean roadblock. Even with an American flag secured to the antenna at the rear of the jeep, the guards weren't taking any chances; they were on extra alert. At the stop point, two soldiers stood by a vehicle. Just north of their position, Joe saw emplaced machine gun positions on either side of the road, facing north.

Foxy's Korean language skills came in handy, and he explained that they were going to see Captain Hui. They exchanged more talk then waved them through.

"Something's up, Joe. They said their regimental commander came through a little while ago to see Captain Hui.

"This could be lucky for us. Nothing like meeting the guy in charge of defending the area all around our Post."

"Let's hope."

They pulled off the trail near a line of jeeps parked outside a large tent. Foxy walked over to the guard, spoke briefly to him. The soldier went inside, then reappeared with Captain Hui and another officer. Foxy approached and saluted both officers while Joe stayed in the jeep.

As Captain Hui introduced him to Colonel Yup, commander of the South Korean 12th Regiment, Joe saw Foxy wince. Odd, he thought.

They talked for a while, and then the three went into Captain Hui's tent. Ten minutes later Foxy came out and walked to the jeep. Joe saw Captain Hui come out and start issuing orders, and soldiers started moving fast.

"What's going on, Foxy?"

"It is our lucky day, Joe. I'll tell you when we get back on the road. We have a lot of work to do."

As they passed through the roadblock, they picked up speed.

Foxy told him about the meeting.

"I wanted you to meet Captain Hui, but it just wasn't the right time."

"Understood. You said this Captain speaks English, is he from the States?"

"No. He was raised in an orphanage in Seoul by Catholic Nuns. They taught him English."

"Oh man, that's awful."

"Yeah. Tough growing up that way."

"No. I didn't mean it that way. I meant being raised by Nuns."

"Nuns? What's wrong with Nuns?"

"That's a story for another day."

"Come on; you started this. Now you've got to tell me."

"I had a few bad experiences with them in grade school. Still angry about it."

"That's a long time to hold a grudge, buddy."

"Some things are just hard to forget."

"That's it! You're going to tell me now." He stopped the jeep. "What was so bad?"

Joe relented. "Well, imagine you're nine years old in a classroom of thirty kids. This big fat Nun catches you doing something you shouldn't be doing like throwing a spitball. She brings you up to the front of the class. Whacks your hands with a pointer stick, then makes you crawl under her desk. She promptly sits, shoves her chair into your face, and lets out a long, quiet, and very smelly fart."

Foxy burst out laughing. "That's got to be one of the weirdest things I've ever heard."

"You should have been under the desk!"

They laughed all the way back to the Post.

As Joe got out, he said, "I'll go hook up the trailer and load the weapons."

"Okay. Be back in a few."

Foxy met Red as he entered the building. "Come in; you need to hear my report."

They sat down across from Rusty. Foxy started right in.

"Captain Hui came out of his tent with a kind of concerned look on his face, Colonel Yup, the 12th Regiment commander followed him. Captain Hui introduced me but introduced me as a colonel in disguise. I was surprised but went with it. I figured the captain knew this colonel wouldn't respect a sergeant, even if I was a US Army sergeant, so I played along, a good thing, too. This colonel was full of himself."

"Think he's capable?" Asked Rusty.

"Yes. He's sharp, a hands-on kind of leader. He also thinks the North is going to attack very soon. Maybe in a day or so. He was there to reposition his regiment, starting with the captain's company. They're pulling back to just north of our Post, across the road at that Ridge. He's spreading the rest of his 1st Battalion across a thinner line at the border. He doesn't think the border is defensible against a severe attack. He also ordered the balance of his 2nd Battalion into positions south of the Stream, just below our Post. That's where we saw that company that was setting up. The other two companies will arrive later this morning. His 3rd Battalion is coming up to take positions two miles further south, just north of the town of Kim Go. They should be set by the end of today."

"What did he say about the weapons?" Red wanted to know.

"Oh, that was funny. When I told him we had two extra bazookas, if he wanted them, he smirked. *No, he said, not those worthless pieces of shit.* I told you he was sharp. I updated him on this new version, he smiled brightly, said, *Yes, Yes, we need them.* He wondered if he could get them today and instruct his men on their use. I said we could."

Red was pleased. "Good. Did he say where he wanted them? What about the mines?"

"He practically jumped out of his uniform when I told him about the mines. Yes, he wants whatever we can give him. Reinforcing both of these new positions would be ideal. One bazooka and mines for each."

"Rusty, we have forty-two rockets for the bazookas. How many do you want to give them?" Red asked.

"Eight per bazooka should do it. That'll leave us plenty." Red agreed.

"Foxy, you'll have to instruct these guys on the new weapon. Red, you'll need to open up the mine depot."

The depot was a mile south of the stream, in high grass, just off the road. It was a shallow covered trench where they had stored over one thousand of their excess mines plus extra ammo and automatic weapons.

Foxy nodded to Rusty, then turned to Red. "I alerted Colonel Yup that you would show up. He's to direct his people to follow you to the depot. They have trucks. He may come along as well."

"Great! I wasn't looking forward to loading those crates."

Rusty thought for a second. "Take Joe with you Foxy. Red, you take Arron."

"Joe should have the bazooka's loaded by now. Red, if you're ready, we'll go with you. I don't think Captain Hui will be ready for some time. I'll get with him later. Besides, I think you may need my language skills."

"Sure would make life easier. I'll get Arron."

Two hours later.

Trucks started pulling into the field in front of the South Korean stream positions. Watching from the south side of the stream, Red pulled up next to Foxy. He and Arron got out and walked over.

"It sure was nice to see other folks work like dogs while you look on," said Arron, remembering unloading those same crates months ago.

"Every dog has his day, Arron," Red said, "let's hope the rest of the day goes as well."

"I finished training the bazooka teams, Red, I'm about ready to go to Captain Hui's position. Did they load trucks for him as well?"

Red nodded. "They were just finishing up when we left. They took the remaining crates. Should be along shortly. So you know, we saw an artillery battery of 105's just pulling in near the depot."

"That's a good spot, they'll be at maximum range to the border, and close by to us if we need some help. I'll talk to Captain Hui about trying to coordinate with them, but it might be a challenge. The good news is that we have the same radio equipment." Foxy said this with a hopeful look on his face.

"Good idea. A little help down the road could be a lifesaver," Red said as he and Arron started walking back to their jeep. "I'll see you back at the Post."

Joe watched as a ROK officer with two truckloads of soldiers arrived on the north side of the stream. The officer started yelling orders as many soldiers jumped out and began unloading crates and opening them. Others, carrying shovels, started digging holes.

Man! Joe thought they aren't wasting any time. Yeah. They knew their lives depend on it!

"Come on, Joe. I need to find Colonel Yup, talk to him about radio coordination. He's the man, not Captain Hui."

Foxy was directed to the command tent. Five minutes later, he emerged with a pleased look on his face.

"We're good to go. Got it covered."

63

It started to rain. "Oh shit!"

"What? It's just a little rain."

"Colonel Yup raised this concern. At the beginning of the rainy season, the rice farmers start planting their seeds. You saw how many dry rice paddies there are around here."

"Yeah, but that's a good thing, isn't it?"

"Well, yes and no. The no part is that the farmers rely on all the local able-bodied men in their district, including their families, to plant. Unfortunately, able-bodied men were either taken to Japan and killed or after the war were conscripted into the South Korean Army. Since then, the Army has allowed the men to take extended leave when the rains start. This could be a disaster for all defensive units."

"Man, that's crazy!"

A few minutes later.

They pulled into Captain Hui's position at the Ridge, just across from the Post, and found the captain directing the placement of troops. He quickly rounded up eight men and came over to Foxy and asked him to instruct these soldiers on the bazooka.

As Foxy started the training, Joe surveyed the surrounding area, wondering if Captain Hui could hold it.

The ridge started just off the main highway and was a rise, that elevated to roughly twenty feet from the sloping field to its front. The area was open due north for about three hundred yards, where a much smaller ridge rose from a tree line. The defensive ridge extended to the east, away from the road, ran about one thousand yards where it met a steep rise of rock that wrapped northward, creating kind of a bowl effect on the open field towards the road.

Joe thought the field was an ideal killing zone.

Behind the ridge, the open field narrowed quickly by the high rock formation from the east, ending with a width of around two hundred feet, paralleling the road for three hundred yards to the front of their Post gate.

As Foxy finished up with the men, trucks loaded with mines pulled into the open field at their front. Captain Hui came over, saluted, and shook Foxy's hand. They spoke briefly. Then Hui went to round up more men to plant the mines.

Foxy came over to Joe. "I saw you studying the layout. What do you think?"

"Damn good, Foxy! It definitely will slow them down. Helps us, too, creates a great blocking position for both of us. Neither of these positions can be flanked from the road. With those low mountains on their east flank, they can't be flanked period. I saw you talking to Captain Hui, what's up?"

"I wasn't supposed to tell him about our hidden explosives on the road, but I did. I figured if he had to abandon the Ridge, he might feel better knowing that we had the road covered. I also told him not to try to come into our compound, for any reason."

"You did the right thing."

"Hope Rusty agrees with you. At least the damned rain has stopped."

Later, in Rusty's office, he agreed that Foxy did the right thing and gave him a big, "Well done."

Alone again, Rusty thought to have that artillery battery in support was an incredible stroke of luck. Thanks to Foxy.

Chapter Eight

Larry leaned into Rusty's office. "Heads up, this is your reminder to call Colonel Nelson in ten minutes."

"Check. Assemble the men by the radio desk. I want them in on this call."

"Got it."

Minutes later, the men gathered around the radio desk except for Snake, Arron, and Joe, who were on security duty.

Rusty announced. "Colonel Nelson is about to give us the latest Intel, so listen up."

Larry made the call, "Colonel Nelson. This is The Post. I have First Sergeant Rusty Fabino on speakerphone for you."

Rusty began, "Colonel, I've got my men here. I thought it best they hear this conversation."

"Hi, Rusty. Great idea. I've got lots of activity reported across the border area. But first, fill me in on what's happening on your end."

Nelson and the men listened as Rusty reported on the day's developments.

"Good work setting up radio contact with the 12th Regiment. Have you made contact with the regiment's 1st Division commander, General Cho U?"

"Not yet. I was going to run through the radio network with the South Korean units tomorrow. Having the division commander's name is a big help."

"Good. I just spoke with Colonel Pyong Oh. As G2 at ROK Army HQ, he's disturbed that his boss doesn't believe an attack is imminent. The General Staff has taken some precautions but won't issue a full alert. This has a lot of dire implications."

Pausing for a second.

"He also says that the rains are imminent, coming fast, perhaps within a day or so. The unit commanders are allowing weekend furloughs to half their men so they can plant rice at home. It's insane! Hard to believe! War is about to start, and half the troops are away planting rice."

"I don't get it either, Colonel, but the Korean Army has made this practice a mandatory thing for a long time. Unless war were underway, a rebellion wouldn't be too far-fetched."

"Good observation. Not much we can do about it anyway." Nelson continued. "Colonel Oh also confirmed that there were three additional plane incursions reported yesterday along the other main attack corridors. They did find that plane you guys shot down, and it had cameras. More bad news, they picked up a refugee this morning who crossed the border near you last night. He told them that he saw lots of artillery about eight miles north of the border. He couldn't describe them, said they looked big and had long barrels. That could describe the 122mm Soviet Corps field gun. We know they gave a lot of these to the NK Army. With a range of twelve miles, that gun would cover the border area and four miles south, which by the way, includes you."

"Thanks for that, Colonel."

"Just passing on the Intel. I don't have any good news. Nobody, with any say, expects an invasion. Follow the plan we laid out eight months ago."

"Do you want me to implement *Tiger*?"

"Affirmative. I told Captain Jordan an hour ago. He'll order the other Posts to do the same."

"I'll commence first thing in the morning."

"Good. Need anything? I can get it on a plane in the morning."

"A few tanks, maybe." The men grunted approval.

Rusty went on, "But seriously, I didn't think the South Korean's would set up a defensive line right next to us. The added support we would get from their artillery makes our position very formidable. We could hold out much longer than we had initially planned. But, and a big but it is, I don't think I have enough men to sustain a long fight. So, if I had a wish, I'd like a few more troops."

The colonel was silent for a moment. "I was thinking along those very same lines this morning when a Leprechaun appeared out of nowhere, like a miracle."

The men all looked at each other, wondered if the colonel was starting to lose it.

"This Leprechaun granted me one wish." He went on. "I said I wanted ten good soldiers that would be willing and able to march into hell."

Nelson laughed. "And you know what!? He granted my wish. Right there on the spot!"

Now the men were getting antsy.

"Is Red in the room with you?"

"Yeah, he's here."

Nelson boomed. "Hey, Red, come over to the mike!"

Surprised, Red jumped up and walked over.

"Hey, Colonel!"

"Do you know a soldier named Tully? A first sergeant in the Rangers?"

"Know him!? Yes, sir, I know him! He's a friend of mine. A good man."

"Glad you feel that way Red because he and his squad are arriving in Seoul tomorrow morning."

Red was ecstatic. "Awe hell, sir, that's fantastic news!" He saw his opportunity and went for it. "You know, Colonel, despite all the things that Rusty has said about you, I never believed any of it.

You're okay in my book." The men laughed, admiring Red for giving the colonel some stuff. The thought of more support helped.

Nelson laughed along with the men. Then he went on.

"I'll have Billy bring them up to you as soon as they land. This may change your needs, Rusty. Anything else you want to add? I only have one plane. It will be pretty full, so keep it reasonable."

"Appreciate your help, sir. It's a big relief knowing we can defend this position the way it should be. Yes, I could use a few things; two more light machine guns, more ammo, and flairs for the 60mm mortar. Oh, the Rangers. Do they have any heavy weapons? Any snipers in the squad?"

"Yes, one is an expert sniper. There's also a BAR specialist."

"Excellent! That'll be good for now, sir. Now Colonel, why don't you go out and meet another Leprechaun, but this time, wish for a regiment." Everybody smiled; they were in good spirits.

When the call ended, the room exploded in noise.

Rusty yelled, "Hey! Hey! Quiet. Listen up!"

The men settled down.

"You just heard that *Tiger* is being implemented. That means we **assume** we are in a state of war with the North Koreans, regardless of official policy by our government. We are going to be attacked and are to prepare to respond and set ourselves up to evacuate immediately. But first, we defend this position for as long as possible. Thanks to reinforcements coming tomorrow, it should be a good deal longer than planned. I'm turning this over to Red to give you specific assignments."

Red stood and said, "Tomorrow's going to be pretty crazy around here. We assemble at 0500. Marty and Tracker! I want you to get the rifle bandoleers set up and the extra Tommy gun ammo out to the forts. Bring more thirty caliber belts out as well. Then dig a firing pit for the 60mm mortar on the southwest side of fort #1."

Marty turned to Tracker and cracked, "Just when I was looking to go on an exercise program, the gods of war answer."

Tracker shot back. "Good thing too! Arron's cooking is starting to show."

Red continued, "Foxy! Take Joe, Arron, and Snake to disperse three of the jeeps first thing, on the way back, stop by that artillery unit and move things along on communications and make sure they have our target maps."

"Larry, tonight after Joe replaces you, check all the jeep radios."

"And Foxy, tell Arron I need to see him as soon as he comes in."

"Roger that."

Rusty called out from his office. "Hey, Red, let's talk."

Red went into the office, sat down.

"Tell me about your Ranger friend."

"I met Tully three months before I came here. But I got to know him pretty well. We spent a few nights on the town with his whole squad. He doesn't brag or like to talk about himself. His men love him. They told me stories about his exploits, how he saved many of their lives by just being smart, tough and ballsy. They saw the awful stuff in Europe. Not that there was much that wasn't, but they were in the worst of the worst. All I can say is he and his men will be a real asset to us."

Rusty nodded. "Good to be Irish and know friendly Leprechauns. Thank you, Colonel Nelson!"

Then said, "Red! Have we covered everything? Have you talked with Arron about provisions for our new arrivals?"

"Covered, Rusty. Arron's been stocking up on extra food and water for the last few days. He knows that he needs more. As you said, it'll be pretty crazy tomorrow, but it'll be organized crazy. You just focus on our defense and escape."

Chapter Nine

"Sergeant!" Colonel Nelson called. The sergeant quickly came to the door.

"Yes, sir?"

"Deliver these orders to the 1st Ranger Company Commander. Then give a copy to First Sergeant Tully of said unit. Deliver the second copy to General Kean at 25th Division HQ. I need it done ASAP."

"ASAP, sir! I'll have Private Davis cover for me."

Colonel Nelson called in his Executive Officer, Major Jarvis, instructed him to issue orders for tomorrow's plane. He was to get the specified weapons and ammo and additional supplies, and alert Kimipo of the morning arrival, then alert Billy to transport the Rangers and supplies to The Post immediately.

Making sure General Walker was in the loop, he then called his Chief of Staff, Colonel Harris, to inform him of this new deployment.

Finally, he called Headquarters Naval Forces Far East, to speak with the commander.

Admiral Turner Joy answered. "Colonel?"

"Sorry to be a pest, Admiral, but it's been two days since I checked in with you about ship deployments. Any changes since we last spoke?"

"Colonel, I'm bored to death over here. Your calls give me something to do, so keep calling." He chuckled. "Wish I could tell you the Fleet just pulled in, but they haven't. As I said two days ago, I've got the Light Cruiser "Juneau" patrolling the Strait between Korea and Japan. The Heavy Cruiser "Rochester" off the east coast of Japan. That's a three-day sail to Korea and finally the carriers "Valley Forge" and the British Light Carrier "Triumph" near Formosa. That's a five-day sail. Of course, I've got some destroyers but nothing else that could help you out. And no. No troop transports within a week's sail."

"Frankly, Admiral, a few days ago, the probability of North Korea invading the South was around 60%. The situation has changed for the worst. I now believe there's a 95% chance. I mean in a day or so. South Korea could be overrun within a week if we can't figure out a way to slow them down…"

"I'm listening, Colonel… tell me what do you need."

"Admiral, is there any possibility of moving those carriers a little closer to Korea? It would cut down our response time considerably. Every day closer would save lives."

"Colonel, I'm on your side. I read the intelligence reports. I know things are getting interesting, but nobody at Supreme Headquarters agrees. But I'll talk to Vice Admiral Struble at 7th Fleet, see if he has any problem moving those carriers closer. I don't see why this would raise any concerns with anybody, shouldn't be a problem. Hell! I'll call him now!"

"Thank you, Admiral."

"Colonel, a man as passionate about his beliefs as you, has my full attention. I'll get back to you."

As he hung up, Nelson sat back and reflected. What else could he do? There was nothing he could do with the Air Force; they were beyond his influence. Damn! He thought. He let his mind wander off this crisis and thought of his wife and two boys. It had been over a year now. Another 4th of July barbecue he would miss. God, he loved cooking out on the lawn. His neighbor, Bob, always put on a great fireworks show. His mood darkened. He pondered the upcoming fireworks show heading his way. Once again, he questioned himself. Had he done enough?

Same Day - Friday – June 23, 1950
Korea - The Far Western Post # 1 – 1830 – (6:30 pm)

In the cramped operations room, seven men gathered around First Sergeant Stu Howard.

"Listen up! Captain Jordan initiated *Tiger* a few minutes ago. You'd have to be deaf, dumb, and blind, not to know an attack is coming. So Colonel Nelson ordered us out before shit starts. I told Captain Jordan we'd evacuate tomorrow night, leave at dark, and head to our first backup location. So get organized. We'll start loading the jeeps in the morning. George and Steve are on security; I'll give them a heads up.

Chapter Ten

Saturday – June 24, 1950
Korea - The Post - 0500 - (5:00 am)

The mess-room was crowded, the aroma of coffee and fresh-baked rolls filled the room. A hot-tray of scrambled eggs was on the table. Arron was cutting vegetables next to a large pot when Rusty walked up. "Nice work, Arron. The eggs look great. You stay up all night?"

"Nah, just my southern hospitality kicking in. Got these eggs yesterday from a local bringing food up to the border troops. I paid twice as much as he would have gotten so that I could get them. He was happy. I wish Tracker had been there to see my sign language skills; he would've been impressed."

"I'm sure."

"I also loaded up on water and food supplies. Got a lot of roots and snagged a rabbit in one of my traps—good thing I did, with the extra folks coming. I thought we might be holed up inside for a while. I got enough food and water for two weeks, but now, maybe just a week. You think that's enough?"

"How the hell did you folks lose the Civil War?"

"Easy! I wasn't born then!" He smiled.

Rusty filled his plate, got coffee, sat down next to Foxy.

"Rusty, I'm worried about my grandma and cousin in Seoul. I want to warn them and give them a chance to get out, go south. My grandma has a brother way down on the tip. He has a boat, fishes for a living. They might have a chance there."

"What are you asking, Foxy?"

"I want your okay to make a detour into Seoul this morning. The last drop off-site is only about eight miles north of the city. It would only take an extra half-hour. It could save their lives."

"Not a good day, Foxy. Tomorrow would be better."

"You said this thing could go down at any time. Today, tonight, who knows?"

"Yeah, I know." Rusty frowned. "You'll have three men with you that I need here. Time is crucial, even if it's only a half-hour. Sorry Foxy, I can't approve this. You need to come right back."

Dejected, Foxy nodded, stood, and walked out of the mess hall.

Red sat down next to Rusty.

"What's up? Foxy looks like he just got assigned to clean the shit pots."

Rusty explained. Red shook his head. "You know, sometimes, I think you lose sight of what's really important."

Rusty reflected. "That's why I have you around, Red, to remind me. But sometimes, I'm forced to choose logic over feelings, to save more lives."

Korea - Same Day -The Post – 0600 - (6:00 am)

Dawn was gray with thick clouds, humid and still. The men had just finished prepping the three jeeps that were going to be concealed, part of the plan of survival, and maybe of future use. The radio antennas were curled down, secured, and their trailers hooked up.

As Foxy got into the lead jeep, he directed the men, "Joe, you got the second, Arron the third."

"How come I'm always last?" Snake griped as he jumped into the last jeep, the one that would take them all home.

Foxy yelled back. "Remember what the Good Book says, *and the last shall be first*."

The men laughed as Snake yelled back. "Foxy, you're so full of shit!"

"Let's get this show moving," He yelled, "Our first stop is four miles away. We'll check in with the ROK artillery unit on our way back."

The convoy drove out, headed south.

Korea – Same Day - Back at the Post

In the storage room, Marty and Tracker were opening crates of ammunition.

"You want to be in charge?" Marty asked Tracker.

"I hate that word. Reminds me of what your cavalry did to my people."

"What in hell are you talking about? What word?"

"Damn, you are dumb. You know, Indian and charge?"

"Holy shit, Tracker! Are you serious!?"

"You're all messed up this morning. Nervous maybe? I was trying to lighten things up. But to answer your question, NO. Only the pale face can be in charge."

"Okay. Since I'm in charge, you load forts one and four. I'll take the other two. Let's load the M-1s first. We need two thousand rounds per man, two men per fort. The bandoliers hold six clips of eight bullets each, so that's forty-eight per sling. Forty slings per man. Eighty slings per fort."

"Marty, you did that in your head?"

"No wonder you guys got whipped," Marty responded.

As they started stocking the forts, Red was the only security guy on duty. He was in the turret, atop the compound.

Korea - Same Day - The first jeep drop off point – 0700 – (7:00 am)

Joe followed Foxy as he turned off. The others waited on the road. Slowly they navigated through low brush, the pathway covered with large trees and bushes. As they neared a high rock ridge, Foxy recognized the spot and pulled to the side of a dense opening. He stopped, signaled Joe to pull in.

They threw branches over the opening, hiding the jeep and trailer. It was invisible from ten feet.

When they got back to the road, Foxy told Joe to stop by the other two jeeps and he got out. Joe wondered what he was doing when Foxy called the men together.

"Here it is. I'm going to disobey Rusty's order. I have no choice. I've got to warn my Grandma in Seoul to leave the city. Rusty said no to my request that getting back ASAP was more important. You know the way to hide the next two jeeps. I think I can do both by not going with you. I'll go directly to Seoul, talk to my Grandma, and pick you up at the last drop off-site. It's only ten minutes from the city. I don't think I'll delay anything. Are you guys good with this?"

Snake said. "This is family, and family always comes first." Everyone nodded but Joe.

"Joe! You okay with it?"

"It's not like you're going AWOL. I'm good with it, on one condition."

"What?"

"I go with you."

Foxy shook his head, "No!"

Arron wasn't having any of it. "You crazy, man? Take him!"

Foxy reconsidered. "You're right!" He jumped into the jeep, looked back at Arron and Snake, and nodded thanks.

Chapter Eleven

Saturday – June 24, 1950
Japan - Yokota Air Base - Flight Line - 0700 - (7:00 am)

Colonel Nelson rode with the airbase flight officer Lieutenant Hanson, whom he had just picked up at the gate entrance.

"Colonel, the plane is just past that C-47 on the right."

"Never saw a plane like that, Lieutenant."

"It's new, a C-119. It's called *the Flying Boxcar*. It just came in yesterday, and it's a good thing too. That C-47 was supposed to fly this morning but had trouble. So this new bird will replace it. We're going to do an evaluation study on this new plane, and we'll start today. I'll tell you it's impressed everyone so far, just by its carrying capacity."

"Funny looking." Colonel Nelson switched his focus to the Rangers standing near the rear of the plane. As they pulled up, the Lieutenant turned to him. "Sir, Colonel Harris told me to inform you that he hopes his *present* is appreciated, said to tell the Rangers good luck, from him and General Walker."

"A *present*?"

"We loaded it earlier. I think you'll like it. The Rangers sure did."

"Stand by Lieutenant; I'll be a few minutes."

He walked up to First Sergeant Tully, who was standing in front of his men.

"Aaaaatension!" First Sergeant Tully yelled and saluted. The men snapped to attention.

Nelson saluted. "At ease!" He shook hands with Tully. "Good morning First Sergeant! I thought I'd come by to make sure these poor bastards actually volunteered." The squad laughed.

Nelson addressed the men. "My name is Colonel Nelson. I'm responsible for you being here today because your first sergeant volunteered all of you to go on this dangerous mission for me. I want to make sure that each of you understands what you're getting into and offer you the opportunity to back out right now if you believe this is not what you signed up for. No recriminations."

He paced slowly in front of them, making eye contact with each man as he spoke.

79

"Let me be clear. You will be coming under fire! Today, tomorrow or the next day, and that's a fact. You will be at the very heart of this battle. You will be three miles from the border when the invasion begins. You'll probably be overrun but escape. Then your survival will rely on your collective skills and cunning. Your job will then be to stop, delay, and cause damage to the enemy until we can get organized over here and respond in force. The small unit you're joining has been there for eight months preparing for this battle. You'll be relying on their local knowledge and tactical abilities."

He stopped and stood still.

"All of you served in the European theater and are **unfamiliar** with the Asian culture. That's unfortunate. If you assume that Koreans are inferior in any way, you will die a surprising death. We underrated the Japs at the very beginning of the war. That was a huge mistake and cost us many lives. The North Korean Army is highly trained and made up of veterans of the Chinese war with Japan and then four more years of intense fighting with the Chinese Nationalists. They are battle-tested and led by trained officers. The Russians have five thousand officer trainers fighting side by side with the North Korean officers."

He paused again for emphasis.

"Last, but not least, the South Korean forces you will be fighting with are not up to our standards of training and experience and have few veterans. They are poorly equipped, have no tanks, few artillery pieces, and no Air Force."

Corporal Mazzoni, the third man from the left, spoke out.

"So far, so good, sir! What's the bad news?" The men cracked up.

"Yes, corporal, there's some. Monsoon season is just about to start."

A Ranger called out, "So what's a little rain, sir?"

Colonel Nelson shook his head.

"Okay, I said my peace. Anyone want to reconsider?"

Total silence.

"Thank you, men. General Walker and Colonel Harris asked that I express their best wishes for your success and safe return."

Nelson turned to Tully. "First Sergeant, do me the honors and introduce me to your men."

"Yes, sir!"

They walked over to the first man on the left.

"Corporal Francois Delcamp. We call him *'Frenchy.'* He's our language specialist. Tell the Colonel how many languages you speak, Frenchy."

"I'm fluent in French and German, sir. Pretty good with Russian, Japanese, and Korean and can get a drink with my Chinese."

"Why aren't you in an Intelligence unit, son? Or, for that matter, why are you still in the Army?"

"Colonel, that's a complicated question, sir, that requires hours of soul searching to answer. Let's just say that I love the Rangers and leave it at that."

Colonel Nelson smiled, shook his hand, and moved to the next man.

"Corporal Kevin Moor, *'Moose,'* he's our BAR man." Colonel Nelson marveled at the size of Moose, 6'4", a solid two-thirty. The match-up was perfect as the Browning Automatic Rifle was a heavy weapon to handle. He shook his hand.

"Corporal Mario Mazzoni. He's our mountain climber and one of my original squad."

"So, you're a comedian, as well?"

"No, sir. After what you just told us, I thought a little levity was in order."

"Tell me a little about your experience in climbing, Corporal."

"I grew up in Seattle, sir. Mount Rainier was a early and often climb for me. I traveled to most of the best mountains in the States; it's an obsession. The Army allowed me to climb some of the best mountains in Europe and Japan. Mount Fuji is one of my favorites."

They moved on.

"Corporal Dean Watkins. Our sniper."

Nelson saw the particular sniper rifle. "Why are you still using the old 1903 Springfield, Corporal?"

"Sir, I've tested every sniper weapon in use today, including our new ones. This rifle is still the best for me. From shot to shot, I can fire a kill shot at six hundred yards every eight seconds, with 90% accuracy."

"Impressive. How many kill shots did you confirm during the war?"

The corporal didn't answer.

Tully did.

"Dean's been with me since the beginning. At Normandy, he fired from the beach to the top as we climbed the one hundred foot cliffs at Pointe Du Hoc as the Jerry's were lobbing grenades down on us. When we got to the top, we counted nineteen bodies with headshots. That was his first day. I can assure you he is outstanding. By the time we got to Bastogne, our squad had guessed he had scored over a hundred kills, but I think it's much more. At that point, we all just stopped trying to count. We were much too busy trying to survive."

Colonel Nelson looked at Corporal Dean Watkins with knowing eyes. "It's a lonely job, Corporal. Few handle it well, but I'm sure you're proud of how many of your men you saved."

"Yes, sir, I am." Corporal Watkins acknowledged the pointed complement.

When finished, he walked back to the front of the line.

"It's been an honor to meet each of you. I wish you luck." He saluted.

"First Sergeant, take me to the plane and show me this *present*."

The men picked up their gear and headed toward the aircraft.

Chapter Twelve

"Hey, Red! How deep do you want this pit?" Marty called up to the turret at the top of the compound.

Red moved to the edge of the rooftop, leaned over its lip, and yelled down, "Marty! Have you never dug a firing pit before? Tracker, what the hell? You saw a lot of action! Are you going all stupid Indian on me!?"

"Watch out, Red! I'll get my war paint on, teach you some manners! We Indians don't dig holes, we sneak around and scalp ignorant white folk."

Red laughed. "Okay, I get it. You two have never dug a mortar pit before. Four feet plus a foot of sandbags. And don't forget the ammo trench off to the side."

Marty yelled back, "We couldn't dig four inches on Okinawa without a pickax! We had to use grenades to make shallow foxholes. Your guidance, oh great one, is always welcomed."

Same Day- On the inbound C-119 to Seoul – 0745 – (7:45 am)

They had just leveled off, cruising at two hundred and fifty mph, about one hour inbound to Seoul. The storage area had seats running the full length on either side of the large open bay in the middle.

The Rangers were seated five per side, starting just behind the pilot's area. The loadmaster told them it was the quietest place, with the wing engines back and above. He was right.

The loadmaster unstrapped his seat harness, and walked past the men to the pilot's door hatch, opened it, and yelled up to the pilot, "Hey, Captain! Can the guys move around?"

"Sure thing, Harry. We're at ten thousand feet, above the cloud cover and should have clear sailing. I'll put the jump signal light on just before we start our descent."

Harry closed the hatch, went over to Tully. "You can walk around if you want. Captain says we should have a smooth flight."

Tully nodded, unstrapped his harness, and stood. His men did the same and wandered around the *Present.*

Corporal Dean Watkins yelled over the sound of the engines to Tully. "I've been staring at this *Present* since we took off, and I can tell you I've seen plenty of jeeps before, but never one with a cannon mounted on it ….and….I've never seen a barrel like that either. It must be seven feet long! You ever see anything like it?"

The other guys nodded. The jeep, with its attached trailer, was lashed down in the center of the cargo hold. They walked around it, inspected the cannon, and examined the sighting mechanism.

Tully let the men tour for a few minutes, then said, "Hey guys, it's not the tenth wonder of the world, but it is a great weapon, and I've fired it. Remember when I went to that Weapons Assessment program, where I met Red? That's where I first saw it. It's called a **Recoilless Rifle**. It saw some action before the end in Europe but never got deployed in any numbers. The cannon fires a 75mm round with a seven thousand-yard range. That's four miles for you dummies."

"Oh, come on, Tully! We're way smarter than we look!" Moose said.

"Hey! Can this thing fire from the jeep?" Ben yelled over.

"On that first remark, you're not. And yes, it was meant to fire from a small mobile platform because of its unique feature, that being it has no recoil. A specially made breach venting system coupled with a unique shell that releases gas on ignition prevents any gun recoil. It's something and very accurate. The only negative, and it's a big one, is that when fired, it creates a tremendous smoke cloud. So if you are exposed, you need to get out fast because everybody will see you."

"Got to be a great anti-tank weapon," Frenchy added.

"That it is! But you need HEAT rounds. I checked the ammo boxes in the trailer, no HEAT, only HE. We can only use it as close-in artillery. Still, it's a deadly addition to our fighting power."

"How about showing us how to load and fire this thing?" Mario asked.

"I will, but first, I want you to get familiar with these map books Colonel Nelson provided." He handed them out. "These are aerial photos with grid marks of the Uijongbu Corridor from the border to Seoul. That's our area. These are your guides for any gun support or rescue mission. Most South Korean units have these."

Mario was looking at all the mountains. "Where're we going?"

"Go to map page three, down to line C where it intersects with grid line two. That's our new home."

Chapter Thirteen

Saturday – June 24, 1950
Korea – Foxy in Seoul - 0800 - (8:00 am)

"I'll wait here."

Foxy walked into the restaurant and saw her. "Grandma! Glad to see you're getting an early start!"

"Grandson! What a surprise! You come to help me today?" She proudly spoke in her broken English.

"I've come to warn you, grandma."

"Warn me? What you mean?"

"Is Suh here with you?"

"He go get vegetables, be right back. What wrong?"

"Listen to me. I have privileged information, secret in fact, that the North will invade us very soon. Maybe today or tomorrow. I want you and Suh to leave Seoul and go to your brother. I mean, right now, today! A train leaves for Pusan at noon. I want you on it."

"Grandson, you in panic. If they attack, we very strong. Have America behind us. We will defeat them!"

"You don't understand! All those stories your President has told you are just that; stories! The North will be in Seoul in three or four days if they attack us now. There will be extreme panic. The railroad bridges will be damaged on the first day, and you won't be able to leave. You'll be trapped. You're right about America being behind us. They are so far behind us that if they do come to our aid, it will be too late for you and the people of Seoul. Maybe the whole country."

Now scared, she reverted to Korean. "I know you know these things, grandson. You are American, and despite what you've told me, you are very important here, and in danger, too. What will happen to you?"

"I'll take care of myself. I'm very well trained and am very good at it. Grandma, I've seen war very close and what it can do to innocent civilians, and I don't want that to happen to you or Suh, so please listen to me. Promise me you will leave today on the noon train."

Thinking about losing her home and maybe her newly discovered grandson, she began to cry. Suh was her life now. She had to keep him safe. Hearing this potential horror brought back those terrible days of losing her husband. No! Not again, she thought. I won't lose my dear Suh. She stopped crying and got hold of herself.

"I will take the noon train with Suh on one condition. You must promise me that you will come to me in Pusan when you're finished, okay? I love you so. I love talking with you, being with you. My daughter did a very good job of raising you. You are a special man, promise me you will stay alive."

It was Foxy's turn to tear-up. He hugged his grandma. "I love you, grandma. I will do my best to meet you in Pusan."

As Foxy walked out, Joe saw him wipe his eyes. He'd felt the love between them at lunch; was struck by it. Now he tried to recall the last time he cried and thought about his grandparents, cold, harsh.... well, there was that one that time... his thoughts drifted.

As they left, they were greeted by civilians who crowded the exit road from Seoul, heading south. Ox carts were packed with furniture, cooking pots, and all manner of supplies. People with poles on their shoulders, with hanging baskets full of their possessions, parents dragging children, and older people carrying what they could.

Joe broke the silence. "How do they know?"

"They just do. Shit, this is going to slow us down. I hope I didn't screw this up."

"Foxy, I just met you, but I already know you're one hell of a guy. I admire you. So let me tell you something. If you hadn't done this, broke the rules and gone to your grandma, I would've lost all respect for you, and besides, you'd've never forgiven yourself. So drop the worry about the timing. You did what you had too."

"Thanks, Joe."

Korea - Same Day - The Post - 0830 - (8:30 am)

"Rusty!" Larry yelled, "I've got the Colonel on the horn!"

"Good morning, sir! At least I hope it's a good morning."

"Yes and no."

"Give me the *yes* stuff first, sir."

"The Rangers are on their way. You'll be impressed. They'll be landing soon and come bearing gifts from General Walker." He filled him in on the Recoilless Rifle jeep.

Rusty was impressed. "General Walker stuck his neck out on this reinforcement, and we appreciate it."

"I do too."

"What's the bad news, sir?"

"There's a lot. We've been monitoring a tremendous increase in North Korean radio traffic from all four Posts. It started hitting the ceiling before midnight and has continued through the morning hours. It's coming from divisional headquarter units going to their individual combat commands. They are mostly coded, but we have broken some of these. Around 0600 this morning, they started transmitting in un-coded, plain language. They reference *Operation Flower* and set a kick-off time of 0400 tomorrow morning. We can only guess that this is the main show. All signs point to it."

"Colonel, it sounds like good news to me. Knowing **when** is pretty crucial, right?"

"Agreed, but I'm not finished. There's an increase in incursion activity all along the line. Most disturbing is an incident in your area. It seems a platoon of South Korean troops guarding the Kaesong railroad bridge over the stream bed near the Post has been compromised."

"Their normal radioman, who reports in every six hours to his captain, was a different guy at the midnight check-in. The captain was told the assigned guy had gotten sick. He didn't think much of it then but started to worry."

"At the six am check-in, he asked to speak with the platoon sergeant and gave a phony name. The radio guy said he wasn't there; he was out on patrol."

"It's Captain Hui's platoon, and he's a smart guy; knows something bad is going down but doesn't have any extra men to send over there. So he called his battalion commander, and he can't spare a company to investigate, so then he kicked it up to his regimental commander, who's reluctant to commit any reserves. Finally, he kicked it to heaven, where it landed on Colonel Pyong Oh's desk."

"Jesus, Colonel, sounds like a Chinese fire drill."

"Yeah! And then it landed on my desk, and now it's landing on yours! Colonel Oh knows the bridge is only four miles from your Post. So he asked me to send a scouting party from the Post to see if it's serious enough for him to move a lot of troops. Nobody believes that the North Koreans would cross the border, go three miles into the South and capture an old wooden railroad bridge that hasn't had any rails on it in five years."

Nelson continued. "I think otherwise. I believe that it's a severe event. So I need you to whip up a little recon operation and check it out. How soon can you do it?"

"I know that bridge; it's strategic. If there is trouble at the bridge, it means a pretty large enemy force must be involved. So here's a stupid question. How could a force that large get through the border?"

"That's the real bad news. The South Korean company guarding the tracks at the border cannot be contacted. Their battalion commander just called a half-hour ago. He sent a squad up to check it out, but he lost touch with them as well."

"Oh shit! Maybe this thing has already started. I'm starting to think that they consider us all assholes, or maybe they just don't care if we know at this point. Either way, they could do a lot of damage if they got behind our lines that quickly, especially if they got that rail line operational. But damn it, right now, I got my guys all over hell and back." He paused, "Let me think for a second."

Thinking out loud, he called out, "Got an idea! This could work. It'll need some luck and some coordination, but I have a plan. I'll have a team over at the bridge within a few hours. I'll get back to you, sir.

"That's why you're there! I knew you'd come up with something. Good luck!"

Chapter Fourteen

"Where the hell are they?"

"Calm down, Snake. See all this traffic? Slowing them down, I'll bet."

"Looks like rats leaving a barn fire."

"Ever had a barn?"

"Yeah, my house."

"You lived in a barn?"

"It was better than the mud hut I was born in. Our family snuck across the Rio Grande when I was seven, found a rancher in Laredo that needed a lot of help but had no money. We bartered for labor, food, and shelter and stayed on. We still live there. The Morgan's are good people. We're considered part of their family."

"I've wondered why you have such a little Mexican accent. How is that? You being raised by Mexican parents."

"Mrs. Morgan was an English teacher at the local school. She started on me as soon as we settled in. She was sensitive to the town's hatred of all things Mexican. They thought we were subhuman, treated us like dirt, so I didn't go to school. She taught me at home."

"She taught you well."

"Yeah, I'm indebted to her. The Army taught me the rest. After Europe, I landed in Japan, took every education class I could get. I love the Japanese. They can't tell a Mexican from an Indian, and they don't care!"

"You'd probably feel different if you'd fought **them** instead of the Nazis."

"You're right. Probably would. I still hate the Jerry's. After the end, I spent three months in Hamburg, spent time with the people before shipping to Japan. I can tell you I think the whole German race is arrogant, heartless, and cruel."

"Hey! Here they come." Arron waved to them.

Korea - Same Day - Same Time - Seoul - Kimipo Airport – 0845 - (8:45 am)

Its ramp already down, Billy pulled the duce and a half up next to the rear of the plane. Earlier, he had hitched up the truck's trailer. Somebody was backing a jeep with a trailer out the aircraft. For the second time in less than ten minutes, Billy did a double-take when he saw the cannon on the jeep. He still hadn't got used to seeing this weird plane yet. Billy's helper, Arnie, arrived with the folk lift.

Looking around the mostly deserted airfield, Tully said, "You must be our ride, sergeant?"

Billy grinned. "Maybe! But first, you got to tell me about this plane and that jeep."

Tully filled him in on the details of both. Billy shook his head in wonder. Man! He thought I hope they make a lot more of these planes! Sure looks easy to unload. But that jeep there...I don't know.

"I usually charge ten bucks a head for pickup and delivery, but I don't get to see many Rangers, so, today, you'll ride first class for free. I'm Billy, First Sergeant."

"Call me, Tully, and thanks for the discount." He didn't like Billy right off.

"Arnie!" Billy yelled. "Load the skids in the trailer, and hurry up!"

The men stretched, gathered their gear, and climbed into the truck.

"Hey Tully, I'll drive the jeep if it's all right with you. Can Ben ride with me?"

"Only if you don't stop to talk to the natives, Frenchy. Read me?"

"Tully, you have no sense of adventure."

Tully shook his head. "Stay right behind us."

He yelled to his men, "Lock and load, be ready for anything."

"I'll ride shotgun with you, Billy."

"Jump in! Firepower is always welcome in my castle."

Korea - Same Day – Same Time - The Post – 0845 - (8:45 am)

After ending his call with Colonel Nelson, Rusty went up to the turret at the top of the compound to see Red.

"I need to bounce this off you. Need a sanity check?"

He filled him in on the situation.

"Oh, hell! Rusty, from one nut case to another, whatever you're thinking can't be any crazier than what I'm thinking."

"Okay....What are you thinking?"

"I say the bridge is occupied and they're restoring it and laying tracks. Hell! They probably have a train there. You can't rebuild that bridge without a lot of material, lumber, and rails, and they sure as hell couldn't have carried all that by hand. That means they'll have at least a full platoon, maybe two platoons."

Red thought about it.

"This is a perfect job for the Rangers. I say we get them over there and destroy that bridge and whatever else they can do to close off the access route. That new gun they brought should be just the ticket."

"God damn! Great minds.....! What about leaving the Post exposed, putting all our resources into this mission?"

"This is our number one mission, Rusty. Delay and destroy. You told me that on my first day here. The Post is well defended, even if we only have six men here. I say we have no choice. We take bold action."

"We make a good team, Red. Thanks!"

Rusty returned to the radio desk.

"Larry, get me Foxy!"

Korea - Same Day - Foxy on the road north – 0900 - (9:00 am)

"Buzzzzzzzz." Arron, his mind far away, was thinking about home when startled back to now, almost jumped out of the jeep when the radio buzzed. He grabbed the receiver, "Arron here."

Arron tapped Foxy on the shoulder, "Pull over! Got Rusty on the horn, he wants you!"

He pulled off the road. "What's up?"

"We've got a situation. Where are you?"

"We're about twenty miles south of the stream."

"Good, I want you to go to that South Korean unit there and stage this mission."

Rusty filled him in and told him what he wanted him to do.

Chapter Fifteen

Saturday – June 24, 1950
Korea - Artillery unit at The Stream – 0945 - (9:45 am)

On the ride up, Foxy detailed the situation to the men. The mood was somber. They were thinking about fighting again after five years. Flashes of horror jarred their minds, wondering if old instincts would keep them alive.

"I'm going to talk to the captain, Arron. Keep an eye out for Billy's truck. He should be along any time. Have him pull in over there." He pointed to a spot just south of the road.

Foxy found the commander of the artillery unit, Captain Kim Soo, who knew about the Americans and Foxy in particular. Captain Hui and his gun controller briefed him. They spoke in Korean.

"I want to make sure we can communicate, Captain. Who is my contact for fire missions?"

"Please call me Kim. I will be your main contact. I direct the battery. You met my fire mission controller, Lieutenant Lee Kun. He told me of your concerns. You have our radio frequency, and we have your Aerial Sector Maps. We are familiar with this entire area. Most of my men are from this area, so we are ready to help you if you need us. We are an excellent unit, trained together for two years now. This is our homeland, and we will fight hard to defend it. I hope this eases your concerns."

"Thank you, Kim. I need to ask you about a situation that's just come up."

"Are you talking about the railroad bridge? Are you going to check it? We, too, are concerned."

"We're going to do more than just check it out. Pretty sure the North has taken it, and we aim to take it back, could need your help."

"My regimental commander, Colonel Yup, called me twenty minutes ago. He told me to do anything I can to support you. So if you need us, call."

"Looks like we'll be badly outnumbered, so I'm planning on it. Let's coordinate a fire mission right now…."

Captain Soo turned to his aid. "Get Lieutenant Kun. I need him here now!"

"Let's check our maps, Foxy." Soo laid out the Sector Map book on his table.

As they studied the maps, Lieutenant Kun entered.

"Sir! You called for me".

"Lee, you know Foxy." The two greeted each other; then, Lee was briefed on what they were planning.

"I will need to turn all the guns and reposition them. That will take about an hour."

Foxy leaned on him. "Any chance you could speed it up?"

"I will start now. I will do my best."

Foxy turned to Captain Kim Soo. "I'm counting on you, Kim."

Soo nodded, "We will not disappoint you."

"One more thing, can I use your tent to brief my men on this mission? I've got a team arriving shortly."

"Only on the condition that I furnish coffee and some food. I will get that organized."

As Foxy started toward the jeep, he saw Billy's truck pull in with the recoilless rifle jeep right behind.

He saw the first sergeant stripes on the Ranger.

"You must be Tully! Welcome to Korea. I'm Foxy." They shook hands.

"What's going on? This isn't the Post, right?"

"Change in plans. We got a mission. Get your men and follow me. Have them leave their gear."

Foxy signaled his men to join them.

"Billy! You too!"

The men entered the tent. A large pot of steaming hot coffee and a platter of assorted pastries greeted them.

"Now, that's what I call hospitality!" Arron beamed. He poured himself a coffee, grabbed a sweet cake, and shoved it into his mouth.

"Go ahead, everybody, help yourselves! Complements of Captain Kim Soo, battery commander of this fine unit." Foxy called out and got himself a coffee, had a sip, and moved to the table in the middle of the tent.

"Listen up everybody; I'm Foxy. For all you new arrivals, Rusty is our Post commander. You'll meet him later. He's put me in charge of this briefing," he checked his watch, "which we'll start in about fifteen minutes. You arrived right in the middle of a bunch of shit, and it's our job to clean it up. Sorry, we don't have more time to get to know each other before we go into combat together, but I think we're professionals and know our jobs. Your Ranger reputation precedes you, so we expect your supernatural abilities to make this easy."

That got a chuckle.

"You'll find that we too, although just regular Army guys, know how to fight."

He briefly filled them in on their experiences.

"Okay, let's go over our mission." He pointed at the map. "We're here. The stream in front of us wanders west for about three miles to the railroad bridge. The rains haven't started yet, so the creek is shallow, maybe two feet deep in the middle. The banks are wide but still dry. They rise gradually and then get steeper closer to the bridge. About three hundred yards from the bridge, the stream takes a full left and forms a narrow peninsula on the right. Then it veers sharply to the right, straight to the bridge. Trees and bushes on either side of the banks will give us good cover. I plan to take our new jeep weapon along the south shore to about four hundred yards from the bridge where the peninsula still blocks the view from the bridge, where it then pulls onto the bank and into the trees. See that small clearing? It's about two hundred and fifty yards from the bridge. We set the jeep with its cannon there. Tully, I want you in charge of this. Pick three men to go with you.

Tully raised his hand.

Foxy said. "Hold your question until I finish."

"Billy will take his truck with the rest of us, proceed along the north bank to the spot where Tully goes into the trees. That's where we disembark and head to the bridge. Billy will take the truck back two hundred yards and wait. How are we doing so far? Tully, what's your question?"

"Yeah," he said, "What if they have lookouts on the peninsular? That's what I'd do."

"Okay, what's your suggestion?"

"See that bend in the stream? It's about a thousand yards before the peninsular. No lookout can see anything behind it. We stop there, send two scouts on foot to eliminate lookouts. It'll add maybe twenty minutes to the mission, but it will ensure we're not spotted. Once we get the all-clear, we'll move into position on both banks."

Foxy looked up, gave him his best poker face, played the pause then broke into a big grin. "I knew you Rangers were sent to save us."

The guys laughed.

"Okay, we'll go with that. I got Snake as one of those scouts. You have a ghost that can do that, Tully?"

"All my men are ghosts, but I have the best of the best, Mario, our climber."

Mario called out. "Never told me that before, Tully. I'm humbled."

"Oh shut up, Mario. You don't even know what that word means." Some hoots.

"Good. Mario, we'll wait until you give us the signal. Snake, you proceed to the rear of whatever is on the tracks and eliminate any guards and stop anybody from leaving. Mario, you're the first scout to guide us as we move to the bridge. You've responsible for taking out any guards."

"Tully, who do you want with the cannon."

"I'll take Moose, Frenchy, and Dean."

"Take Joe, too. He can take the south end of the bridge and seal that exit."

"Where the hell am I going to sit in the damn jeep?" Joe gripped.

"You'll look awfully good riding that cannon, Miss America!" Arron shot back.

Foxy thought the men's laughter was a good sign, anxiety prevention.

"Tully, here's a walky-talky, it's set to my frequency. We'll coordinate the attack. You take out the bridge with the first shot fired, make sure it goes down. You're on your own after that. At the same time, I'll be calling in an artillery barrage from this unit we're visiting. Questions?"

Foxy waited. Dead silence.

"Alright! Unhitch the trailers! Stow your gear next to them. Tully, take four or five rounds for the Cannon or whatever you can fit in the jeep."

On his way out, Arron snagged the last cake on the tray. Got to find out how to make these beauties, he thought.

Chapter Sixteen

Saturday – June 24, 1950
Korea - The Post – 1000 - (10:00 am)

Tracker leaned against the dirt wall.

"Marty, I think we're done. I, for one, am done in. It sure was a lot easier when we had that backhoe to dig these trenches." Both were drenched in sweat.

Marty was more philosophical. "Feels good working hard, lets you know you're still alive."

"You must have grown up in a lovely home, easy life because working hard sucks."

"Well, yeah. I did have a nice family, but my parents worked hard. My Dad was a machinist; my Mom had a part-time waitress job. We lived in a tenement house in a bad neighborhood, but it was okay with me."

Marty changed the subject. "You know we have an Indian reservation in Connecticut, not twenty miles from Bridgeport, and I never met an Indian before you."

"I feel special somehow."

"Stop clowning, I mean it. As a kid, I always fantasized about Cowboys and Indians, you know, like Tom Mix. It must have been wonderful growing up in the middle of pristine wilderness."

"You're right, Marty. I grew up in beautiful natural surroundings, big mountains, lush valleys, and right on Lake Superior. It doesn't get any better than that. But I came to hate it. Around eight or so, I realized I was condemned to survival. Brutal winters, very hard work just to get by. No one knows poverty unless they come from it. Being an Indian in a white world is depressing, like looking up from a bottomless pit, not seeing any light. The Army saved me, taught me how to accept the white world, then embrace it. I'll never go back home. My home is here, with the Army."

"I never considered that." He paused, thinking what to say next. It came to him.

"Okay, next time you're the boss."

Red shouted down from the roof. "Stop goddamn loafing! Get off your asses and finish. Lay some timber boards in that pit, so it doesn't turn into a fucking swamp when it rains, and get the mortar set up.... Let's go!"

Chapter Seventeen

Billy stopped the truck. The men got out and took defensive positions.

"Let's go, Snake!" Mario crouched low to Snakes left, as they both moved quickly and got to the peninsula in five minutes, then slowed down; Snake moved to the west side.

Mario sniffed the air. Cigarette smoke! He traced the aroma to a soldier sitting next to a fallen tree, looking down the stream in the direction of the truck. Binoculars hung from a branch, his rifle at his side, leaning on the tree. Oh, yeah, Mario thought, his cap had a big red star on the front; the uniform is not olive drab, definitely North Korean.

The guy was dead in a move and a slash.

Mario swept the area to the end of the peninsular. Suddenly, Snake appeared at his side, startling him.

"Jesus! Snake. You almost made me shit myself! See anybody?"

"Sorry. But yeah, just inside on the north edge, there are a few sentries. I'll scout the rear approach to the rail line. You get the men." They both moved out, Mario, in a run.

With the all-clear from Mario, the men climbed back in the truck. Foxy radioed Tully, gave the go-ahead, and added, "Hey, Tully! That was a good call about the scouts. Saved our asses!"

The truck moved forward to its position and Tully to his.

"You do kinda look like a prom queen riding a float at homecoming." Moose cracked, as Joe straddled the Rifle cannon on the jeep as it sped ahead to the turn at the spot on the bank.

Joe smirked. "I do have the best view, paid extra for it!"

As Billy turned the empty truck around, Tully drove the jeep onto the top of the bank. The going slowed, as the bushes and trees got denser. Finding a small clearing with good cover but also enough access for line-of-sight firing at the bridge, he pulled in close to the bushes and stopped. The men jumped out and started setting up the Rifle cannon.

Tully grabbed the binoculars, took his first view.

"Oh shit!"

Frenchy was standing behind him. "What's wrong?"

"Just a lot more folks at the party than we thought, and.... there's a train."

Tully snatched up the walky-talky. "Foxy! Do you read! Foxy!"

"Affirmative!"

"I have a clear view, lots of troops, and a train. It's a locomotive and two flatbeds. I see twenty or so workers on and around the bridge. Another twenty around the train, but I can't see the rear area. I make it two full platoons, guessing sixty men."

"Okay, let's speed this up. I'll call in the artillery five minutes after we open up. Where are you on your set-up?"

"I'll be ready with the rifle in five minutes, but I want to give Joe ten minutes to get in position. So let's set our time, I got 1050 now; we fire at 1100."

"Check! I'll order a barrage at 11:05. They'll do a full six-shot mission, should take about two minutes, then we'll see how it goes from there."

"Clear here. Good luck!"

Knowing he had ten minutes to get to the south side of the bridge, Joe scrambled to the edge of the dense bush cover in five. Sandbag protected machine gun positions were on either side of the approach track, three men in each. He couldn't see any more nearby but saw troops working on the bridge twenty feet away. Okay, he said to himself, maybe five at the bridge are a danger, but they're working, not carrying weapons. I'm parallel to the machine guns, so this should be easy. He checked his watch, three minutes to go.

Foxy talked to each man down the line, told them the new plan. They started moving closer, wanted to be twenty-five yards from the train. They still had good brush cover. He told Mario to find Snake at the rear, to stay with him, protect his back. He checked his watch, three minutes to go.

"Moose! Tully said, "Get your "BAR" set up to rake the bridge from that position," he pointed to a downed tree at the bank's edge. "Dean, move further west, so you have a better view down the train, do your thing, officer's first." Dean had already worked on his sniper scope and was ready to make the final adjustments when he moved out. Tully checked his watch, three minutes to go.

"Tully, all four rounds are armed," Frenchy whispered. Tully was lining up his next target after he took the first shot. Got it, he said to himself and reset the weapon's site to the primary target.

"One minute to go," Tully said.

"Not to be stupid or anything, but is this Rifle loud." Frenchy looked at him straight-faced.

Tully smiled, then checked his watch and fired. All hell broke loose.

The shell hit the bridge three feet up from the base on the north side. The bridge disintegrated, collapsing in on itself; the north side fell first. Soldiers were flying through the air, some trying to jump clear, most were being hurdled off, others fell straight down, getting buried in the debris.

Joe jumped out, running and firing directly at the closest machine gun position. Then the far one. Finished with them, he got to the train track and headed to the bridge, changing a clip as he ran. He took out two men, then two more that tried to run. Another guy was sitting, stunned; he had his hand on his pistol still in its holster. Joe yelled, signaled for him to raise his hands. But he panicked and pulled his gun out. Joe fired a burst and cursed him as he fell. "You idiot!" Damn it, he thought, could've taken you, prisoner.

Moose fired at the rail edge as soldiers ran away from the bridge. The explosions made a big cloud of dust that hampered his view of the train area.

Dean had a better view and was scoring.

Tully fired the Rifle cannon again. The front of the locomotive exploded.

Near the train, Foxy's men kept up their continuous barrage of fire from the bushes, causing mayhem. It was up close and personal for the surprised North Korean troops, many died standing.

As he made his kills, Foxy turned robotic, detached, his thoughts bizarre. How many men could you kill per minute? The M-1 can fire forty rounds per minute in five clips. My "Tommy" fires six hundred rounds per minute, but I got to change magazines after thirty rounds. That takes four seconds. What does that equal? How many dead in one minute?

He forced himself to stop his diversional calculations, got focused. The enemy started firing back and getting organized. But they had taken a terrible beating at the beginning. Dead bodies lay all along the track.

Foxy checked his watch. One minute left. Time to move back.

The shooting died down as the men redeployed further back into the trees.

Suddenly, the screaming sound of an incoming shell pierced the air. It exploded behind the train.

Foxy snatched the radio. "Captain Kim! Down twenty feet! Fire for effect!"

The next sound he heard reminded him of a horror movie, screaming ghosts. He couldn't remember the movie's name. Funny, he thought. Just then, six shells detonated two hundred feet in front of him. Everything in his view disappeared in smoke and dust. The concussion knocked him flat to the ground. In the seconds that followed, he heard nothing.

He buried his head in the dirt, put his hands over his ears, waited for another visit from hell, and hoped Kim stayed on target. He did. They came. Five more volleys. It lasted two minutes. How many kills, he wondered? Then said out loud to nobody. "I hope every last mother fucker!"

The next thing he heard was the eerie silence. Still feeling groggy, he pulled himself up.

He got up, yelled, "Move in, men!"

They had seen destruction like this before, five years ago. A bad memory, now real again.

Body parts everywhere, ugly smells of blood, and shit - utter destruction.

"See if anyone survived, we need a prisoner." He called out to no one in particular. The men moved amongst the bodies and mangled rail cars. Foxy started looking for an officer near the locomotive, moved north amongst the wreckage, and found the body not far away. He searched the body for papers, anything that might reveal something about their plan, found nothing. Standing up, he noticed the bullet hole in his forehead. Nice shot, he thought, wondering if it was that Ranger sniper.

On the South Side of the Bridge

Joe moved out from his cover after the barrage lifted, went to the bridge end to check for any survivors in the river bed. Still hazy from the debris, he saw bodies entwined with the wooden bridge parts. He heard screaming to his left, just below him. A soldier, lying flat on the bank, had a large wood piece of bridge sticking out of his stomach.

Joe watched him scream, wondered how long before he died. On Okinawa, sometimes he'd put a wounded Jap out of his misery, other times not. A lot depended on how the firefight went, or his mood. But mostly there wasn't much mercy shown by anybody on that stinking island. But this is a new war, a new enemy. Best start it off right, he concluded. He put a short burst into the enemy soldier.

On the Northside of the bridge

"Foxy! Got a prisoner!" Yelled Snake as he approached, coming down the mangled tracks with an NK soldier in front of him, hands held behind his head.

"Good job, Snake. Looks like a sergeant."

"That's what saved him. He was taking a leak in the bushes before the firing began. I saw his stripes and decided he might be more valuable alive, so I just knocked him out."

"Where's Mario?"

"He's watching the north tracks."

Foxy approached the man and took a hard look at him, then spoke to him in Korean.

"Are you a sergeant?" The man nodded.

"How long have you been in?" The man looked at him, not expecting this question. "Ten years."

"You've seen a lot, perhaps you've heard that Americans don't kill prisoners?" He nodded, yes.

"Good, now let me tell you that that's generally true, but circumstances sometimes change this rule, and this is one of those circumstances." The man stiffened.

"So here's the deal, I need information from you, and I need it right now. If you decide not to cooperate or if you provide false information, I'll kill you. Cooperate with us, and you live. Understand?"

The man, visibly shaken now, started sweating. He didn't answer.

"Where is the rest of your unit?" Now trembling, the man looked down.

Foxy reached across his chest, unsnapped the .45 holster, and drew the automatic pistol, and clicked back the arming lever.

"I won't ask again, where is the rest of your unit?"

"I don't want to die; please don't kill me." He pleaded.

Foxy raised his 45.

"NO, no, don't," he stammered. "I'll tell you. My company was divided. Two platoons were here, one platoon was at the Pass, a mile north of here, and the other stayed at the border where we crossed."

"Where is your captain?"

"He was here; I think he's dead."

"Who's in charge of the Pass unit?"

"A lieutenant, the company's executive officer."

"Do you know his orders?"

"I was there when the captain told him to come to our aid if he heard gunfire."

"Thank you, sergeant; you'll live another day."

Foxy yelled out, "I want a skirmish line at the rear of the train, NOW!"

The men responded immediately.

Foxy picked up the radio, "Captain Kim, Captain Kim, come back, urgent!" He radioed.

"The North Korean's have a platoon at the Pass, a mile north of us. I'm pretty sure they're on their way here. I need you to stop them."

Kim responded. "Hold on, let me check the map." A few seconds later, "I got it. I suggest we do a rolling barrage from the Pass on down the track line to just in front of your position."

"Yes, that's perfect. Do you have air burst shells?"

"Oh yes, that's what I'm planning to use, I'll mix them in with HE, we'll fire in".....he thought, "In three minutes."

"Sooner is better!"

Foxy changed frequencies and called Tully, explaining what's going on and telling him to come over.

"Leave Joe there, for now. I still want someone at our rear. Give him your radio."

Foxy turned to Snake, "You've in charge of our prisoner, stay here and watch the middle of our line. If he moves funny, shoot him." He repeated this in Korean so that the sergeant would understand, nodded that he did.

Foxy ran up the rail line, at least what was left of it anyway, jumping over bodies, railcar parts, and shell holes. He found decent cover to the side of the track in a ditch, then grabbed his binoculars and scanned up the rail track and the woods on either side. Seeing ahead maybe one hundred yards until there was a bend where the rail bed turned and disappeared. Thick bushes and trees lay beyond thirty feet on either side of the tracks. Looking at the iron rails, he thought that these were all new, and the North Korean's must've laid these last night. His mind went off again. How many rails can you lay with X-men in Y time? He stopped, I think I'm going crazy, he mumbled to himself.

Boom, boom, continuous sounds of distant explosions echoed down the tracks. The men lay in cover along the skirmish line, five feet apart, waiting.

In a few minutes, the explosions were slightly visible, but the sounds were loud. The airbursts were going off in the treetops, which was maybe thirty feet high.

Ben Holt was one of the new Rangers, a real tough guy, grew up in South Philly near the shipyards. He was always in trouble. Ben was one of those guys that could start a fight in an empty room. An original Tully guy from the very start at Normandy. He was prone in a ditch near the tracks, next to a big rock, watching the artillery barrage as it moved closer to their skirmish line. Then the air bursts became visible going off in the trees.

He stared as if he saw a ghost and froze. His mind flashed back to that day, five years ago, that horrible day in the Hurtgen Forest. Wounded by tree splinters, he just laid where he fell and watched the explosions high up in the treetops. He was in so much pain, but hearing his friends scream made him crazy. Finally, he passed out. When he awoke, he crawled among his dead friends; half his platoon was unrecognizable.

He remembered it like it was today, he's watching it, again, now. He started to shake and then put his head down and silently cried.

Jeff shared the opposite side of the rock in the trench with Ben, and he too remembered that day. "Hey Ben, I sure hope to God we never go through a day like we had in that forest in Germany, remember those air bursts?"

No response.

Jeff turned to see Ben with his face buried in his arm. "Ben, you okay?"

Ben turned his face away. "Yeah, yeah, just got a damn bug in my eye."

The shelling stopped.

Quiet.

Nothing. Three minutes. Nothing. Foxy called Captain Kim. "Bring your next barrage set down towards us, one hundred feet, but don't fire until I tell you. We got no movement."

"That's a good sign, no?"

"It's an excellent sign, Captain, yes it is. I'll call you back if I need that barrage. I'll call you back anyway, as soon as I know what's going on."

Ten minutes now, Foxy thought. I've got to find out. Maybe they went back to the Pass?

"Mario, Snake, I need you to scout that line to the Pass. You come running if you meet any resistance. Okay?"

"Right!" They moved out, Mario on the left, Snake took the right. Both hugged the edge of the bushes.

Foxy checked his watch, six minutes since they left. Nothing.

Mario was looking up at the high hills up ahead. They got much higher off to the left. Still, he thought, no climbing here.

Bang! He jumped flat into the bushes at the rifle shot. Bang! Where are they? Is Snake all right?

He heard Snake's Tommy gun rapid fire. He got up into a crouch and raised his weapon. Can't see anything, shit!

Another burst from Snake's gun.

"Mario! You Okay!?" Snake yelled

"Yeah, is it clear?"

"Just a bunch of bodies here, come on, you've got to see this."

He walked towards where he thought Snake was. He still couldn't see him in the bushes. As he crossed the tracks, he saw a body about ten yards away, near the bush edge. Further on up the line, he saw a lot of bodies strewn around. Snake was twenty feet ahead, leaning against a tree.

"The bastard must've been the lone survivor. See that splinter in his leg. Would've bled out pretty soon; I think that's why he missed you."

Mario now noticed the trees and the surrounding forest floor. I've seen this before, he said to himself. Must be fifteen bodies all around them.

"Those airbursts are just so from hell!" Snake said, echoing Mario's thoughts.

"Looks like they were all bunched up, running. Let's check further up the rail. I saw a bunch of bodies on the tracks."

Back at the Skirmish Line

Two shots rang out, then two bursts of the distinctive fire of the Tommy gun.

Foxy wanted to run to them, send two more guys, do something. He had to wait. He knew that. Damn it! He wished they had a walkie-talkie.

On the South Side of the Bridge

Joe was sitting on a fallen tree just off the tracks. He was staring at the three dead soldiers in front of him. He thought about that song that was playing in the mess room on his first day at the Post, "Far Away Places." Funny! Who knew I would end up in Korea, staring at three dead guys that I just killed, on my first day. I wonder where they came from? Why did they have to die today? How come I don't give a damn about them? Why don't I want to go home? Where is home? He took a deep breath. I'm alive! They're dead. Good!

111

At The Skirmish Line

He saw them come into view as they cleared the bend. They were walking casually. "Thank God!" Foxy called out. The other guys cheered.

They briefed Foxy.

"Captain Kim, Captain Kim." He radioed and got an immediate response.

"Cancel that last firing order. Your guys did a great job! You wiped out that NK platoon. The rail line is clear to the Pass. Have you spoken to your Commander about what's going on?"

"Are you kidding? The fan is flying, as you Americans say." Foxy laughed.

Foxy called Rusty, briefed him on the mission, and made a suggestion about the jeep with the cannon. Rusty agreed.

When they cleared the Stream, Foxy directed them to the ROK Artillery unit, where they turned over the prisoner and gave the jeep and ammo to Captain Kim as a gift for his help at the bridge. It was well-received. Kim thought Captain Hui, up at the Ridge, could put it to better use. He'd bring it to him later or tomorrow.

Rusty and Foxy knew they had no use for it at the Post because if the invasion did happen, the jeep rifle cannon wouldn't last through the first hour. But they sure got good use out of it today.

Chapter Eighteen

Saturday – June 24, 1950
Korea - The Post -1200 - (Noon)

The men piled out of the vehicles. Arron was the first one to approach Rusty, standing near the entrance door.

"I'm gonna rustle up some grub; the guys are starving." He moved right past Rusty.

Red came out, saw Tully. "Hey, you Irish son of a bitch!"

"Red! You have a hell of a way of introducing Korea." They embraced, slapped each other on the back.

"Damn, good to see you. Hear tell you did a fine job mucking up the Commies plans. Hey, meet Rusty", He turned to Rusty, "Rusty...Tully." They shook hands.

"Your reputation has preceded you.....and now confirmed. You Rangers are the badest of the bad. I'm glad you're here. Welcome to your new home, although I'm not sure for how long."

"What can I say? We Rangers like a grand entrance!"

"On a more serious note, that Foxy of yours planned the whole operation. He should be an officer. He'd go far; your men did a great job. If I didn't know any better, I'd think they were Rangers." More smiles.

"Red, get the Rangers settled, then give'em a tour." Said Rusty.

"Sure. I'll show'em where to unload their stuff first." He went off to direct the men.

Foxy came over to Rusty and wanted to know what was the follow-up. Tully joined the two.

"What's going on, Rusty? Have you talked to anyone?"

"Are you kidding! I've been on the horn since you kicked that hornet's nest. The ROK's moved quickly, bringing a company to the Bridge area and another to the Pass. Also, they sent a reserve battalion to chase that NK platoon from the hole at the border. But major shit's flying everywhere back home, as well."

"What do you mean?" Foxy frowned.

"Where to begin? Let's see. MacArthur thinks the South Koreans are trying to start a war with the North. South Korea's President thinks his Army is staging a rebellion, and nobody believes this is a prelude to an invasion. Go figure!"

"Sounds like a FUBAR to me. Things never change, do they?" Said Tully.

"Yeah, *fucked up beyond all recognition* describes it perfectly. At least we in the trenches know what's going on." Rusty commented and turned to the door.

"Come inside, Tully, let me show you around."

As he turned, "Hey, Rusty, hey." He heard Billy yelling at him as he approached.

"I need to get the hell out of here!" Billy shouted, "I'm a Goddamn maintenance guy; I don't belong in a war zone. That battle at the Bridge scared the crap out of me!"

Rusty stood rigid, looked carefully at Billy. He got angry, but then, he thought that Billy maybe hadn't been in the War and probably had never seen combat, but still.

"Billy, listen to me. You're in the US Army, in a foreign country that's about to go to war. If you think the enemy gives a shit about what you do, you're in for a big surprise. My advice to you is to find that rifle that was issued to you, load it and take some target practice; because, in a very short time, your life will depend on how well you can defend yourself. Now get your ass out of here!"

Chapter Nineteen

"If I get one more call accusing me of trying to start a war, I swear, I'll hurt someone. Is everybody insane?"

"General, how does that saying go, *we are surrounded by the enemy, and they are us* or something like that?"

"That's a fair assessment, Colonel, as always, you clarify the situation."

"So, I take it General MacArthur doesn't believe the Action Report I sent to his office?"

"Heck! I don't know anything. Korea's President, Syngman Rhee, keeps calling him, wondering if America is part of a conspiracy to overthrow him. This guy is a real whack job. How did we ever get involved with him?"

"I don't know General. That's way above my pay grade. But we do have a few questionable relationships around the world. He's just one of them."

"You're right. Let's focus on what we can do. By the way, that was a hell of a job your guys did."

"According to these new Rangers, 100 to 14 just were terrible odds for the enemy; they never had a chance."

Walker smiled, "Those Rangers! That was a good call you made."

"It helps to have a Leprechaun on your side."

The General looked at him sideways but didn't say anything, then asked, "Any suggestions at this point?"

"If nobody sees this incident for what it is, then I'm at a loss. We have done everything in our power to be ready to respond and have responded very well, I might add. Beyond that, we'll just have to wait until there's a lot of blood on the ground."

"Hell!" General Walker uncharacteristically cursed.

Korea - The Post – Same Time - 1400 - (2:00 pm)

Most everybody had eaten when Mario entered the mess room. Smells good, he thought. Arron's reheated stew was on the burner. Snake was alone, sitting at a table, facing the back wall. Mario filled a bowl and sat down across from him.

"This is good stuff."

"We eat like kings here." Snake acknowledged.

"That was some really seriously great work, Snake. We make a good team. I officially make you an honorary Ranger."

"What the hell does that mean?" Snake said skeptically.

"It means that you can join in on any of our bar fights, come to our fancy Balls and fraternize with our women."

Snake laughed. "I'm honored!"

Mario got serious, "Did you go home after the War?"

Snake also got serious. "Yeah!"

"How come you returned to the Army?"

"I never left the Army. When we shipped out of Europe, Galveston was where we were going to be mustered out. My CO asked me if I wanted to make a career of the Army. I was a sergeant then, too, seen a lot of action; he thought I would be good for the Army. He was a lifer, had fifteen years in."

"Anyway, I told him I wasn't sure. I hadn't even thought about it. He said for me to think it over, gave me a two-week pass. He said that if I didn't want a career, he would muster me out right away."

"When I got off the bus in Laredo, I knew something was wrong. It was bizarre. The ranch where my family lives is a five-mile walk from the bus station. I had to walk a mile through town, then four miles down a dirt road outside of town. But as I went through town, I felt like I had landed on another planet. No debris, no bodies, just regular people walking around, looking happy. It freaked me out. What was worse was how they looked at me. Mostly white folks lived in town, and they don't like Mexicans, even if they're American soldiers. My family was glad to see me, but I couldn't talk to them about anything normal. I didn't understand them anymore. I didn't fit in, couldn't. I left in a week. Decided the Army was where I belonged. Been happy with that decision ever since."

Mario looked at Snake, wondered if Snake had ever told anybody before. He knew he hadn't.

"You and me, Snake, we'll take the world on together." Both looked at each other with knowing eyes and smiled.

Korea - Outside the Post – Same time - 1400 - (2:00 pm)

"What's your thoughts on positioning my men, Red?"

"I think we don't want anybody in the building when the shelling starts. If they hit us with big stuff and hit the building, the concussion inside will kill, safer out here in the trenches."

"I agree, the trenches are deep, excellent protection."

"I'll leave it up to you to set your men, you're as good as they get, Tully."

"Some say other things about me, but I'll stick with yours. I'll put Tony and Manny on the mortar, they've worked together before. I'll position the rest between the forts. I don't think we need the machine guns we brought along, save them for later if we need them. I can see where our main threat will come from, so I'll set Dean, our sniper, and Moose, our BAR man accordingly."

117

"I've got two new super bazookas in the building. Don't need them now, but later we may if tanks show up."

"I brought two more, plus ammo. Big improvement over the old ones I heard."

"I forgot you brought more. We'll store them as well. Go do your thing, then let's get settled in, maybe we'll see something tonight."

Korea - Outside the Post – 1630 – (4:30 pm)

Dean had walked the entire trench two times so far, scouting his shooting angles, anticipating where an enemy sniper would set up his position. He did not doubt that they would have a sniper or two, maybe more. Dean knew the Russians believed in snipers, and they'd been training the NKs for two years. He'd studied how the Russian snipers were trained. They were pretty damn good, he thought, but not as good as the British.

Stopping next to fort # 1, on the northeast side of the fort, Dean said yes, this is an excellent spot to start. He took out his binoculars and scanned the tree line on either side of the open field. They'd want to take out the machine gun crews and have clear sights on both from either side. He adjusted a few sandbags to get a better shot from each side and worked at trying to remain as concealed as possible. His was a game of stealth and wits. The looser paid a high price.

Red came out from the fort. "Hey, don't mess with my sandbags!"

Arron followed. "Don't pay any attention to Red; he's just screwing with you, Dean."

"I figured. Either that or he's been severely affected by some Korean curse. Do they practice voodoo here?"

Red laughed, "Yeah, it's called the Korean *mombo jumbo*. Everybody affected goes round and round, holding their balls, seeking relief. You haven't been here long enough to know that there aren't many women around. Our team's been in country for eight months and are starting to show some early signs, so be careful."

"I know the signs. Somewhere around Bastogne, my foxhole buddy started looking attractive."

They laughed.

"Hey Dean, I hear you're a pretty good sniper. I'm curious. How'd you pick that up? I'm not aware of any Army sniper program. Did the Rangers have one?"

"No, Arron, they didn't, I learned my real skills by chance. I was always a good shot with a rifle, grew up in South Dakota on a small ranch near the Black Hills, did a lot of hunting. I shipped over to England in early '43; we were just setting up then. I had a lot of time on my hands, so I spent my days at the firing range and my nights at the local Pub. I met a Brit sergeant at the Pub playing darts, he was a trainer, taught snipers. One thing led to another, and I got into his program. It was an eight-month deal. Towards the end of the program, the men started calling me Dead Eye Dick. Thought it was a British thing."

"I heard the Russian snipers did a job on the Nazis. They must have had a real good program?" Red asked.

"They sure did and still do. Russia has the largest sniper program in the world. Their training was excellent but not as good as the Brits. The thing was, the Brits turned out maybe a thousand snipers during the War, disbursed them into all kinds of units, never concentrating them, used them as an afterthought in any battle."

"The Russians, however, went full bore, trained over 25,000 snipers, and used them as units. They probably single-handedly won the *Battle of Stalingrad*. I think their top five shooters scored all their kills in that fight."

"I had no idea." Arron reflected, "Did they actually keep score?"

"Oh yeah, that was a big deal for them. They were rewarded for each kill."

"So how many Nazis did their top guy get?" Arron was really into it now.

"That's the point. The Russians didn't just have one guy; they had five. All scored around five hundred confirmed kills."

"Holy shit!" Arron was surprised. "That's twenty-five hundred men, killed by five guys. That's a whole regiment!"

"Right. Pretty impressive, isn't it. Somehow our military leaders haven't seen the benefit yet. We still don't have a formal program."

"Anyway, that's why I feel fortunate to have gotten into the Brits program; it was extraordinary. I saved a lot of men because of that training."

"That's what I heard," Red said, "I hope you've kept up with your training."

Dean nodded sheepishly.

Chapter Twenty

Rusty had just finished going over the large map in his office with Tully and Red. They discussed the fallback plans and reviewed the current disposition of the ROK units in their area.

"I smell Arron's cooking, and I missed lunch. You guys hungry?" Tully and Rusty nodded.

"Then, let's get some; we all missed lunch." Said Red.

Arron was alone in the mess room when the three men entered.

"I don't know what you're cooking, but it sure smells good." Arron answered Red, "Well, it could be our last meal." Then added quickly, "I'm not entirely sure what I'm making! Just throwing in anything that might go bad."

"Oh, come on, Arron, are you feeling all right? You sound gloomy?"

"I don't like being boxed in, feeling like a turkey in a farm pen on Thanksgiving morning."

"Looks like somebody got your feathers already." Rusty was trying to be light.

Grim, Arron turned to Rusty, "You sure *this* is going to work out?"

This has got to be going through everybody's mind, Rusty thought. Hell, it's going through mine. Red and Tully looked at Rusty, waiting for his answer.

"I wish I had the answer, Arron, but I don't. We've all been in tight spots before. Good planning, being smart, and brave, and with great luck, have allowed us to survive so far. I'm hoping that continues. Now, let me at that concoction you're making, I'm starving."

Arron stood still for a moment, then said, "Take from the right pot, I'm still working on the other one," then continued to chop up fixings on the side table.

The three men sat at the table, started funneling the stew into their mouths. They were hungry.

Rusty put his spoon down. "I've been thinking; it's going to start raining any minute now, and it's only 1700 hours. Maybe we should bring in half the men now and let them eat, give them a few hours to take a break, say four hours, then switch to the other half. That would bring us to 0100. All our intelligence says that things shouldn't start happening until 0400. What do you guys think?"

"Shoulds don't count," Red started, "but I think it's a good idea."

"Me too," Tully added.

"Hey, Arron," Rusty called over, "Do you have enough in that pot for eight guys?"

"Oh yeah, plenty, this second pot will be ready in about an hour or so."

"Great! Where did you get all the meat in the stew, what is it, rabbit? Good, by the way."

"Of course, it's good. Well, there's some rabbit and other stuff that you really don't want to know about."

"You didn't have all this meat yesterday, did you?"

"This morning was shopping day at the bridge. That artillery barrage killed more than the North Korean's at the train. Birds, rabbits, and things were all over. I can't stand seeing good food go to waste."

Red got up, "I think I've heard enough. I'll get the men."

Japan – Tokyo –Colonel Nelson and CIA Station Chief Howie Meeting At a Patio Bar near the American Embassy – 1730 – (5:30 pm)

"Thanks for coming, Jim. They've got single malt, your favorite. I usually come here after a bad day."

"I'm up to my eyeballs in crazy stuff, Howie. This meeting better be important."

The waiter came over; they both ordered Scotch, neat.

"I've got to finalize my report on the Korean situation a little later and wanted to talk to you before I send it out."

"Aren't the reports I'm sending you enough?"

"Your reports are excellent but tend to be extremely factual, as they should be, but they seem to be missing something. Maybe your passion, so I want to hear it personally."

"Since when does the CIA give a shit about passion?"

"Stop it, Jim. You know damn well what I mean."

"I take it you're not getting anything from the Russians."

"Andrey has been keeping a tight lip these last few days. Nothing is coming out or going into the Embassy. That itself is very suspicious, but it's only speculation. Besides, I don't think the KGB has anything to contribute on whether or not the North Koreans have gotten a go-ahead from Stalin."

"How many years have I known you, Howie? Let's see, since '44 when you were an OSS operative in France. That's a long time, bud, so I'm thinking, something is on your mind that you're not telling me, so what is it?"

"You really shouldn't have turned down Donavan's offer, you know. You would have been my boss now or even more probably, The Director. So it comes down to politics, as always. Right now, nobody in this administration wants to hear anything about Korea or potential war of any kind. So my neck and my **boss's n**eck are in an awkward position."

"You've been in tough spots before, what's different now?"

"Before, I had Donavan watching my back on political shit, but he's not around anymore. Our new Director is more political than a Company man, and he doesn't want to rock the boat. Can you believe that he ordered my boss, the Director of the Intelligence Directorate, to tone down his reports on North Korean incendiary activity, and to focus on any hard facts about their intentions?"

"I'm confused, doesn't the intelligence that my guys have gathered qualify as hard facts?"

"Only if they know it was intercepted by our **secret equipment and by your men running it.**"

"Oh shit!"

The waiter arrived with their drinks. They toasted and took a healthy swig.

Howie leaned in, spoke softly.

"Only a few people know about your Posts and less about this equipment, and we want to keep it that way. So I've cultivated the story of highly placed spies along the border who are compiling this info. The Director wants solid facts, not reports from spies of questionable reliability. Hence, my dilemma."

"Are you saying The Director of the CIA has no knowledge of the Posts or this secret electronic equipment?"

"Yes, only my boss and I know about this at the CIA."

"You play a strange game, my friend."

"Fieldwork makes you paranoid. It's my only defense."

"Okay, so here we are on what I think is the eve of the invasion, be it tonight, tomorrow or the next day, and you want me to do what? Wait! Before you answer that. I want you to know that General MacArthur only recently found out about the Posts. Not their real role. Only that number one, Eighth Army, meaning me, took it upon itself to have *observation/training* Posts in Korea that doesn't report to KMAG or for that matter anyone else, and number two, the Post's men recently *interfered* in domestic matters."

"Would that last part relate to your guys destroying a North Korean Company at that bridge?"

"You're quick."

"I guess you're telling me your neck is in a noose too."

"Howie, I always knew you'd go far. Look, you and I have done everything we can to alert the people who could do something to either prevent this war or materially alter the balance of power, but nobody cares. Let's face it; we don't know if they will invade, maybe this whole troop movement thing is just a massive exercise by the North to intimidate and shift our focus away from China and Formosa. I'm going by my gut and feel you are as well, but it's our job to do what we've done so far, read the tea leaves and make an intelligent guess about what comes next. We've done that. Now we have to prepare ourselves for the next act."

"Man! You must like theater?"

"Someone much smarter than I once said that *All Life is a Stage.* Do you want me to do something, Howie?"

"You've addressed my problem in your own incredible way. You're so goddamned smart; you make mere mortals like me seem trivial."

"Oh please! You of all people know some of the screw-ups I caused. Come on, Howie, I'm just like you, a man trying hard to do what's right."

"You're wrong, Jim. You're not like anyone else, you've got a gift."

Korea - The Post – 2000 - (8:00 pm)

The first shift was lying around, some in bunks, and some on the floor. Earlier, Arron had heated two big pots of water, and a few had taken hot showers and shaved.

Rusty came over to the radio desk. "Larry, I need to talk to Colonel Nelson." Larry picked up the cable phone and made the connection almost immediately.

125

"Colonel, we're not picking up anything here. It's weird, nothing at all. How about you? Are you getting anything from anybody else?"

"Nothing, Rusty. The North Koreans have gone silent. I think this definitely confirms our time frame."

"Sure sounds like it. We're ready as we'll ever be. What about the other posts?"

"The other posts are starting to reposition now; they wanted to wait till dark. I spoke to Captain Jordan a few minutes ago. He was just about to pull out and be set in a few hours. He's moving to the hide-out between the valleys, the one you guys established, the Peak he said. I understand it's a great location."

"Oh yeah, it is, we spent a lot of time setting that up. I hope he remembers all the bobby traps we set."

"He mentioned that. He said it's starting to rain. Not hard, but steady. Is it raining there?"

"Yeah, started a few hours ago, same, light but steady. Not sure if that's good or bad, but thinking it might slow them down."

"Oh, I wanted you to know that we got to the bottom of what happened to that missing ROK Company at the border. It seems the company captain moved the unit about two miles away just after dark, then disappeared. Nice to know we have some traitors among our friends."

"Thanks for the heads up, Colonel. Just what I wanted to hear."

"Think its gonna be a long night. I'll be here. Good luck."

Chapter Twenty One

"Get the secret equipment into the bunker room and destroy it! We leave in fifteen minutes." Sergeant Stan Miller nodded to his first sergeant and left.

Stan had sent off the last transmission a half-hour ago, but he shouldn't have bothered. Nothing had been intercepted since much earlier. Strange, he thought, from crazy traffic to none. I'm glad we're leaving now, this thing is real; they're coming.

Odd that these were his babies from the beginning. He had taken good care of getting them to work flawlessly and liked working with them. Now it was time to end their usefulness.

There would be more toys to play with, so, with a shrug, he started destroying the essential elements by hand, then with a hammer. Finally, he placed all the pieces in a pit and poured acid over everything.

First Sergeant Stu Howard looked around his command post for the last time, called over to Sergeant George Lopez, "George, you're the last one out. Check the charges and give us five minutes before you detonate. We'll be at the gate."

The gate was one hundred yards down the path from the entrance to their compound. They'd loaded the four jeeps and the trailers for their move to the fallback position farther south. It was pitch black, and it was raining.

Ensuring proper connections for the umpteenth time, George moved from one C-4 charge to another. The last check, he thought. Exiting the post, he jumped into a sandbagged foxhole ten yards off to the right of the entrance, connected the wire to the detonator, and pulled up the plunger.

His watch read one minute to go.

Then all hell broke loose at the front of the compound as gunfire erupted with machine gun and concentrated rifle fire from close in at the gate. He ducked, looked back to see tracers coming from the entire front of the compound. He heard Stu shout something, then an explosion by the jeeps. He saw his men slumped over two of the vehicles. Another jeep flipped and was on fire, the other, empty with men lying still on the ground. Firing stopped. He heard a Korean yelling something.

He froze. Dead. They're all gone. He yelled into his head *MOVE! Or you're dead too!* Without thinking, he pushed the plunger down. Post #1 blew up. Some debris flew about, but not much, and it didn't go far, because it was intended just to destroy stuff we didn't want our enemy to find.

George grabbed the detonator, crawled out of his foxhole into the darkness, moving toward the southwest corner of the perimeter where he hadn't seen any firing. He hoped that the enemy hadn't seen the light from his flashlight as he made his final inspection in the building. Did they believe it was a timed detonation? They'd find the wire leading to his foxhole soon enough. He scrambled faster.

The initial shock and fear subsided and felt he was far enough away to believe he might get out without attracting any attention. He composed himself and started thinking. Shocked, finally accepting what happened, he felt his loss of his friends, his family. They were all gone! How did this happen? We had a full platoon of South Korean troops dug in around our compound. Did they turn on us? Are they dead, too? Shit! I'll probably never know. Reflecting on this for a second, he concluded that the one thing I do know is that from now on, buddy, you trust no one.

George had many talents. But he was particularly adept at night fighting and avoiding detection. These skills were finely honed during three years of crawling around the Philippines, evading capture by the Japs. He and one other from his squad lived to welcome MacArthur's forces, and earn himself a Silver Star for his exploits in disrupting the enemy behind their lines.

The North Korean invasion was beginning and nobody knew it but him. Soon this area will be overrun, he thought, got to move fast and stay invisible. Then George felt his rage begin to grow.

Korea - Same Time - The Post – 2100 - (9:00 pm)

Foxy grabbed his slicker and Tommy gun. Damn that break went quick, he thought. Heading to his post at fort # 2, he put his slicker on then relieved Tracker. When it started raining earlier, they had made a small lean-to on one side of the fort; it was a good shelter for breaks and scanning this side of the perimeter. He decided he'd do three minutes on, three off. That'll keep my mind busy.

A little later, he was looking through his binoculars out to the wire when he heard, **in Korean,** "Good evening." From behind him.

"What the hell!" Yelled Foxy, almost jumping out of his skin.

"Sorry, didn't mean to scare you." Frenchy was standing a foot away.

"Damn, if you didn't! You Rangers get off sneaking up on people? I thought for sure you were a Commie!"

"Yeah, we do that. Second nature, I guess."

"Well, don't do it again! Could've killed you!"

"Don't think so, but hey, I'm sorry. I won't do that again. Room for two in your shelter?"

"I'm ready for a break, sit down. How well do you speak Korean?"

In Korean, Frenchy responded, "I speak Korean pretty good for only two weeks of lessons, don't you think?"

129

"Wow, you speak very well. Two weeks? Come on! How did you do that?"

"Seriously, I'm blessed with this gift. All I need is to be around someone for a little while, and I'll pick up their language. I don't understand how or why, but I do. My ability is greatly enhanced when a woman's involved, and there usually is. No. There's always a woman involved."

Foxy laughed. "How did you learn Korean? You just got here!"

"That's a long story."

"I think we have some time," he checked his watch, "I know we have time. I need to look around every three minutes or so, just keep talking; I'm all ears."

"The Korean thing began in Japan last fall. Everybody knew I could speak a couple of languages. I know French, German, Russian, and Japanese. So I was often called upon by the base officers when a situation came up requiring a translator."

"So one day I get called to the hospital. They got an old guy who's pretty sick; they think he's Korean. I don't speak Korean, so I try my other languages. I hit it on Russian. Turns out, he's from Mongolia, raised in the Russian language and became a fisherman, got married and settled in a Russian hamlet with his relatives in Korea. Then he was taken by the Japanese as a slave. Now he was dying and knew it. He told me he had lived in Inchon with his wife and daughter. Asked me if I would go to them and say what happened to him and tell them that he loved them and wished them happiness. He gave me the details of where they lived and died the next day. I connected with this guy, felt awful about his fate. I decided I'd do something good in my life. So I went to Inchon on leave for two weeks. Boy, that was two weeks I'll never forget. I found his house. His daughter answered the door; she was twenty years old and gorgeous. His wife died long ago, and the daughter was grateful I came to send on his wishes. She tended this small farm right off the bay. It was November and cold, but we managed to stay warm. When I left, I could carry on most conversations in Korean. I got much better with the language after I met a Chinese

Korean lady. That's when I picked up a good deal of Chinese as well."

"You're a piece of work, Frenchy! Gotta say that I'm relieved that someone else can speak Korean. That's going to come in handy."

Chapter Twenty Two

Slickers on, the rest of the second shift had finished and left the building. The rain was heavy now. Rusty was inside and alone with Larry, monitoring the radio.

"I don't think you'll need to stay with the radio tonight, Larry. I put Ray at # 3 fort with Snake. I want you at # 4 fort to take my spot. You'll be close enough to the radio in case we need to communicate with Colonel Nelson. Both Foxy and Frenchy have a walkie-talkie and know the frequency settings of our ROK artillery unit coordinator."

"You're not staying in here, are you?"

"No, too dangerous, I'll be roaming the trenches."

Rusty was the last to leave the building a few minutes later. For the next hour, he made the rounds, stopping and talking with each man, making small talk, complaining about the rain, asking about home. He knew everybody hated waiting, the fear of the unknown. The only known was that death was approaching. Was today my day?

He found Red by fort #1. "I think we should take the machine guns off the fixed poles, lay'em flat in the fort."

"I should have thought of that. Get right on it."

Rusty moved on down the trench, met Tully.

"Hey, Tully, lovely night, huh?"

"It's not cold, and it's not snowing! God, I hate the cold. So by those standards, I'm feeling pretty good about tonight."

"Got to agree with that."

"Weren't you at Bastogne?" Tully was curious.

"Yeah. We were on the northern flank, about five miles to the west. Doing the same thing we're doing now. Waiting, except it was zero and snowing. Of course, we didn't have to wait too long. Then we played hide and seek with the Krauts for a week, in the snow. You were there, right? I think Red told me."

"Oh yeah. It's funny how you remember things. Our unit was beaten up pretty good in the Hurtgen Forest, so we were pulled out of the line and were sent south of Bastogne to rest, as a reserve for Patton's Third Army. I was so relieved at that time because there was not much going on there. We got cleaned up, had hot food, rested, and got some replacements. That lasted three days. Then the Nazi's spoiled everything. Because we were on the southern flank, closest to Bastogne, we ended up joining the point division for that hundred-mile walk. Fighting and walking, walking and fighting, zero degrees and snowing. What fun."

Rusty paused at the memory. "I almost lost my mind in those few dark days. But I kept thinking it was a God joke he was playing on me. You know, Hell wasn't hot at all, and it didn't happen after you died."

"Heard that the Nazi attack was a real horror story. Do you still think that way about God?"

"Don't know what to think, I survived, I only see crazy."

"You at least saved those airborne guys, the 82nd, and the 101st, wasn't it?"

"Yeah, yeah. And did we ever get any thanks from them? No! Fact is, they said we wasted our trip and that they had the Jerry's right where they wanted them. Can you believe that!"

"I never heard that story before, but now I have a new respect for the Airborne; they sound a bit like we Rangers." They both laughed.

Korea - The Post – 0330 - (3:30 am)

Drip, drip, drip, the only sound Joe heard as he sat on a box of ammo, in the trench, fifteen feet from Moose. The rain dripped from his helmet onto his slicker, right next to his ears. His Tommy gun was on the floor, flat against the trench wall. His M-1, the weapon of choice for any distance shooting, was under his slicker. He couldn't imagine the enemy getting close enough to use the Tommy gun. The minefield in his front will be a slaughterhouse, he thought.

He started talking to himself in his mind. What did those Greek Spartans think when they saw the thousands of....hell, I can't remember who they were. Right! That's the point, Joe, isn't it, a few against many. Who gets remembered? Why? For the greater good? Glory? Because they were there? Were they insane? Maybe on some sort of drug? No, none of that, he finally thought. They had a mission! Just like our boys at the Alamo. Just like my men trying to take that ridge so long ago. I have a purpose now! He said out loud. "I will do my duty!" That jarred him from his thoughts; he looked around. Moose was looking at him.

"You alright?"

"Just dozed off, had a nightmare."

Chapter Twenty Three

Sunday – June 25, 1950,   Day of the Invasion
Korea - The Post – 0400 - (4:00 am)

Like a distant thunderstorm, rumbling and remote bursts in the clouds signaled the beginning of War. The invasion began.

Rusty watched the shells flash in the border area, three miles north. A torrent of flashes, like lightning striking every few seconds. The sounds followed, muted but loud enough. He couldn't think of anything to say. Happy New Year? His mind raced. He ran to Larry.

"Get a message off to Colonel Nelson. Tell him we think the invasion has started. Heavy artillery is impacting at the border, and it's heading our way. Larry! Be quick and then get the hell out of the building."

The border barrage lasted five minutes, then started moving towards them. As it crept closer, the ground started vibrating. It came rolling towards them like hot lava spewing down a volcano.

There is nothing more frightening on earth than watching death coming directly at you. The men of the Post watched the slow walk of the monster detonations as it approached.

Knowing what's coming to an experienced veteran, who is willing to stand his ground in the face of horror, is something few understand. Still, this experience, facing a rolling oncoming barrage, was terrifying, even more for them.

When the wave finally reached their outer perimeter wire, four-hundred yards out, they hunkered down and got as small as they could in their trenches, faces covered, some with faces in the mud against the wall, all holding their ears tight against the terrible sounds to come.

Then explosions engulfed them; the noise was unimaginable; shock waves shuddered through their bodies. Dirt rained on them, then concrete. It seemed to last an eternity. The explosions! The shock waves! The ear-piercing sounds!

But it was only a brief visit to hell, as it passed and continued south. Nobody moved.

"Medic! Medic!" Ben yelled.

Marty was fifty feet away, started crouch running up the trench. Smoke was thick in the air. The barrage was now hitting south of them, but still close and thunderous.

"It's Sam! He's buried! Help me get him out!" Ben yelled to Marty.

Marty couldn't see. It was pitch black. Ben was on his knees, pushing dirt away from Sam with his hands. Marty got his flashlight out and shined it on Sam. He wished he hadn't.

Ben freaked and fell backward. "He's mush! Oh, my God! Sam!"

"I'm sorry." Marty turned off the light.

Tully pulled Ben away, brought him over to fort # 2. Frenchy came over to them; they were all trying to calm Ben. Sam was his close friend.

Rusty stayed with Marty, assessing the scene. "Marty, get Larry and bring Sam into the building and put him in my office, cover him."

Rusty felt around the outside of the trench, where the shell had exploded. Damn it, he thought, the shell landed about three feet in front of his position. He realized that if Sam had been prone in the trench, as he should have been, he'd probably still be alive. Must have been kneeling or sitting. Stupid! No, not stupid, just a mistake. Leaning against the back of the trench wall, flashes of his past mistakes came to mind. Have I made any here?

Shells continued to fall nearby but randomly.

He did a full inspection around the entire trench, told Larry to get a message off to Colonel Nelson, and then met up with Red when he finished.

"Hard to believe that we survived all that with only one causality."

"Good planning always saves lives, Rusty! You did a great job."

"Tell me that tomorrow, if you can."

"I think the building took a few hits. You been inside?"

"No, been too busy. Anything yet at the wire?"

"Nothing. Can only see shadows when shell flashes go off, it's black as spades tonight."

Larry found Rusty with Red. "We've got a problem. Can't get through to Nelson. The VLF cone must be down. The ground cable isn't working either. I think the line's cut."

Red took charge.

"Not much we can do about it now. Maybe we can fix the break in the cable line, after daybreak, see if you can see a crater where the line comes up from the road. We'll talk about trying to repair it later."

Korea - The Post – 0445 - (4:45 am)

Heavy shelling was going on just to the north and east of the Post. The South Korean company, to their east, on the Ridge, was taking hell.

Dean had second thoughts about his first choice of spots. He now anticipated that the action would start in the woods on the west side of the fort, he moved from the northeast side of fort # 1 to the northwest side and came upon Moose, who moved into the second spot Dean had prepared earlier.

"Moose, get the hell out of my spot!" He whispered.

"What! Get lost! You got a reservation?"

"Yeah, I do! Didn't you notice that this is a prepared position? Well, I prepared it."

"So what! Get another one! Besides, what's so special about this one anyway?"

"Moose, you know, you can be a real jerk sometimes. Of course, it's special, that's why I picked it, gives me a full view of those woods, that's why, oh shit! Moose, just get out of here."

"OK, okay, I'll move. Don't have to get all bent out of shape."

Dean settled in, built his little rain tent over his rifle scope, and started scanning the tree line. He saw movement, lots of shadows back in the trees. Soldiers came into view, crawling towards the wire. He heard somebody come close to him.

"See anything Dean?" Rusty whispered.

"Troops at the wire. Looks like a lot more in the woods," he whispered back, not taking his eyes from the scope.

"I'll be right back." Rusty ran the trench to Foxy in fort # 2.

"We've got troops in the woods," he pointed. "See if you can get Captain Kim to get some rounds in there. Now! Foxy."

"Captain Kim, Captain Kim!" He called. Nothing. He repeated the call, then again. A Korean answered.

"This is Foxy at the Post; I need a fire mission, where's Captain Kim?"

"He's dead. I'm Lieutenant Lee Kun. We met yesterday."

"Can you help us?"

"Cannot now, we're supporting Captain Hui's company at the Ridge, near you. But maybe later we can help. We got walloped, lost two guns, and a lot of men."

"I've tried getting hold of Captain Hui, is he on another frequency?"

"No, we only have a hard-line connection; his portable doesn't work. Good luck."

He broke the connection.

"We're on our own, Rusty."

Rusty took off and ran, head down, past fort #1, and turned into the mortar pit. "Hey! We got troops in those woods," he pointed. "I want you to put a big load along that tree line until I tell you to stop. I'll get Tully over here to help, but start getting those rounds ready now. When you hear the first shot, you start. Okay?"

"Got it, Rusty, we're ready."

He found Tully, filled him in, then went back to Dean.

"What's happening?"

"More troops moving up, got an officer near the front directing his men, looks like any moment now."

"Can you take out the officer?"

"What kind of question is that? Of course, I can!"

"Then do it!"

"Now?"

"What kind of question is that?"

Bang.

Then the thump of the mortar started. Boom, boom, the mortar rounds impacted quickly. The enemy troops were waiting for their dead officer to give the order. Finally, after the second mortar round hit, the woods lit up with enemy fire.

"You've got the range, keep firing!" Tully yelled. Manny was dropping the shells into the tube, and Tony was adjusting the angle of fire, traversing down along the tree line. Tully decided to get more rounds out of the side storage trench. The mortar kept sounding, thump...... thump........thump.

The defense line opened fire into the wire and tree line, focusing on the flashes of gunfire and occasional bright light from shell explosions. Bullets were impacting the building behind them. They heard the whomp sound in front of them as the sandbags they were behind, did their job. The # 1 and # 3 forts were silent, as Rusty didn't want to expose their locations to fire this early in the fight.

The dull thud of smaller explosions started as the enemy entered the minefield. A bugle sounded. Firing erupted from the northeast woods. The defense line returned fire, but not fort # 2, they waited.

The explosions in the minefields grew in intensity. The mortar kept firing. Having made a full run down the tree line, it now traversed back and reversed its detonations and also walked them in by a 100 yards.

"Tully! How's your ammo?" Rusty yelled into the pit.

"At this rate, we'll be out in about ten minutes!"

"I'll get you more. Switch your fire to the northeast woods. They're massing there."

Tully wasn't sure about that order; it seemed pretty intense right in front of him.

"Manny!" He had to yell now, "Shift to the northeast woods, we're getting more ammo, so keep it going."

They turned the mortar, set the range, and fired it in two minutes. "Nice work," Tully yelled to them over the noise.

"Down ten yards, right five, fire for effect." The woods came alive with bursts.

Marty was the first to arrive with a box of ammo. "Hear you guys are abusing your ammo budget." He dropped the box at the back of the pit. "I'll be back."

Red was watching the enemy advance through the minefield on the west side, holy shit, he thought; they're clearing the minefield by stepping on them. Must be a new Commie strategy, clever, he mused. Wait till they get inside the second wire, he thought with a glint in his eye.

The attack on the eastside didn't get far, maybe forty feet into the first wire. The mine explosions had had the desired effect. A bugle sounded a different cadence this time. Firing slowed as the enemy withdrew back into the tree line.

The west side was different. The enemy had been persistent, ignoring the horror amongst them; they kept advancing until there were no more men. Back on the westside, Dean was scanning the tree line, searching for snipers now. He was feeling pretty good about this fight as he had scored four officers so far. Idiots, he thought, waving their pistols, arrogant, just like the Germans.

Rusty came over to Tully. "No causalities, everybody's good so far."

"They underestimated us; don't think they know what's here. Probably shell us next, use their mortars."

"I agree. Think they got cocky after running over the border positions so quickly. Still hear the South Korean's on the Ridge; they're putting up a good fight too."

"Let's hope these guys continue being unaware of what's here."

Ferrumph, ferrumph! The sound of enemy mortar shells firing caused everyone to drop deep into the trenches. Explosions began all around them, then continuous now, on the building, and along the trench line at its front, every ten seconds, kaboom!.

Rusty crawled up to Tully. "Go back to your mortar team! See if you can gauge the enemy's mortar location. My guess is they're just north of the western edge of the woods in the open field. Give them a little something to think about."

As Tully entered the mortar pit, he banged his leg on a box of ammo. "Damnit!" Tony and Manny were setting up rounds for the next mission from the four boxes in the pit; two more were at the trench opening. Marty had been steadily going back and forth from the storage area in the building.

"Let's try to find their mortars," Tully announced. "Come here! I want to show you. Form a tent." They opened their slickers, pulled them over their heads, and crouched together to keep the flashlight glare hidden. "Our first guess is that they're in this corner, here," he pointed at their Arial Sector map. "That's about a thousand yards. From the volume of incoming, I'm thinking maybe three or four teams. Let's break up their party, spread the rounds around a bit."

Tully got his binoculars focused but couldn't see a thing. Dawn wasn't coming for another hour, at least. Still, pitch black. It didn't matter; he wouldn't be able to see them if it was noon, the position they're going to fire on was on the backside of the tree line.

Tony and Manny repositioned the weapon, Tony adjusted the elevation for a thousand yards, the weapons max range, guessed at the line of fire, and Manny started dropping in rounds. They fired eight shells, a box load. The enemy mortar fire became very sporadic for the next hour.

Dawn broke slowly, with the rain steady and dark clouds masking the rising sun. The background noise of artillery fire and closer explosions was constant. Rifle fire from the enemy line was continuous.

Dean steadily moved around the line, to the westside, then over to the eastern side. He varied his firing positions and timing. At each new spot, he set his rain tent, scoped his target and fired, moving on to the next place he thought was a random location. One-shot, one score. As the morning light started to clarify the battlefield, he stopped.

Rusty scanned the western approach. Jesus, he thought, must be over a hundred bodies, and they didn't even make the second wire. Red, beautiful work, he thought. Now the fun begins. They'll soon see what we've got.

Enemy machine guns opened up, raking both the west and northern sides of their lines. The Post positions stayed low and didn't return fire.

Red saw it first through his firing port in the fort's sandbags. "You shitheads," he yelled. "Okay, so you want to play rough." He screamed as he left the fort and ran low along the trench. As he passed fort # 4, he yelled to Marty. "Follow me! I need your help!"

They exited the building with Marty carrying a backpack filled with rockets, and Red shouldering the newly arrived bazooka. "Red, I'm getting pretty tired of being a mule around here." Red didn't know Marty had been running a restocking operation for the mortar ammo earlier. "Shut up, Marty."

Red got to Dean's sniper spot just south of fort # 1 and stopped.

"Move over!" Red forgot Dean's name. "I'm taking your spot." Red lifted the bazooka and placed it into a prone position through Dean's opening in the sandbags. Tully saw Red with the bazooka, then ran into the fort next to him to see what was going on.

"Give me a round, Marty!"

Marty loaded the rocket, set the firing mechanism, and tapped Red on his helmet.

Tully looked through the firing port in the fort, "Oh shit!" He dove to the floor as the howitzer in the tree line opened fire. The front of the fort exploded. Tully screamed.

Red fired right after the howitzer did. The howitzer disappeared in a cloud of smoke and flying pieces.

Red entered the fort, looked at Tully on the floor, then at the front wall of the fort. "No damage! Thank you, you stinking Nazis!"

Tully got off of the floor, looked around, amazed. "Jesus, you sure as hell stopped the attack, but why the hell aren't we dead? That was a direct hit! And what do the damn Nazis have to do with anything?"

"It's a long story."

Tully was a little shaky, so he sat on a sandbag. "I think we have a few moments. I want to know what saved my life."

"All right." Red put down his bazooka and turned to Marty. "You think you could rustle us up some coffee?"

"I've become a manservant now! What's next?"

Red found a sandbag near Tully, sat down, took off his helmet.

"When you went through Europe, I'm sure you met the Jerrys 88 gun emplacements a million times. Did you ever stop to really look at'em?"

Tully thought for a moment, "Well, I sure as hell remember trying to take those bastards but look at em. No. I never did."

"Well, my curiosity got the better of me one time because I couldn't believe what I had just witnessed. My unit was supporting a Tank company. We were attacking a small town, just inside Germany, crossing a small field when this 88 opened up from woods edge about three hundred yards away. It took the lead tank right out. Three tanks fired at that position almost immediately, with two direct hits. That 88 fired three more rounds, killing all three. Two more tanks engaged, scoring more hits. They died too. Finally, we got air support that knocked out that gun. I had to see this

143

position. I couldn't believe anything could take such punishment and survive. That's how I came to discover how to build these."

"You copied their design? Just looks like sandbags to me."

"Oh, no. You only see sandbags because those inside are used to replace damaged bags outside, they cover up the concrete."

"The whole fort is constructed like a retainer wall, made of concrete with steel rebar embedded vertically and horizontally. The wall of the forts is five feet thick, and four and a half feet high made in a horseshoe shape. Sandbags placed in front of the concrete are three feet thick and five and a half feet high."

"No, shit!" Tully smiled, "So we're snug as a bug in a rug!

Maybe you shouldn't be so hard on the Nazis in the future."

"You, of all people saying that! Even if it did save your life, I still hate those bastards."

Chapter Twenty Four

"General Walker on the line, Colonel."

"Good morning, General."

"Nothing good about this morning so far, Colonel. Have you heard anything more?"

"No, sir. Only that message received from the Post at 0400, stating they were coming under heavy artillery attack. Earlier, Colonel Oh confirmed this information in a flash alert that he received from one of his border units nearby, that they were being shelled at the same time. He also confirmed that a general attack was underway along his entire border. That was two hours ago. But that's it. We've received no additional information."

"This is disturbing. Why don't you have more intelligence from your Posts, it's been three hours now."

"The only reason they haven't checked in again is that they can't. I haven't been able to contact Colonel Oh or his office either, but I'll keep trying. As soon as I get more Intel, you'll be the first to know."

"Colonel, I want you to be aware that General MacArthur's staff is still very pissed off by that "Bridge" action your men took yesterday. I got calls last night and this morning wanting your head. General Almond intends to start a "court-martial" against you for disregarding standing orders of no interference into domestic issues, not to mention that nobody even knew you were in Korea. Thank God we have that authorized memo somebody signed. Remember it said to observe and train, that's all we're supposed to do. This may all be a moot subject now, I hope."

"I'm only concerned about my men right now, General. Anybody over at MacArthur's HQ worried about this attack?"

"Colonel, they don't think this is serious. Besides, they have every confidence the South Korean Army is capable of crushing any incursion."

"Right! Sir!"

Korea - The Post -0900 – (9:00 am)

"What do you make of this, Tully?" Rusty was kneeling at the mortar pit opening. Not much had been going on for two hours.

"Probably found out Rangers are here. Maybe they're thinking about surrendering." Rusty grunted. "Seriously, I believe that they're waiting for more help, men, maybe armor or planes. Could be any or all."

"Not planes. This weather will keep them out of action." Said Rusty.

"This rain does have some benefits. Tanks then, or armored vehicles. More men won't change the situation for them."

"Arron went into the building about a half-hour ago to get some stew heated up. Why don't you start sending three men at a time to chow down? Oh, and have each man bring back two rounds of bazooka rockets. Bring them to the north trench."

"Good idea."

Rusty went over to fort # 1. Red was looking out the view hole, "Hey, nice shot, Red! How'd it feel? The new bazooka, I mean."

"I shot plenty of the older model, and this is much better, faster rocket to target, nice kick but straight back, doesn't jerk; I was surprised. It's extremely accurate."

"I like good news. Got an idea; need your feedback." Rusty leaned in, told him his plan.

Arron came up to them. "What a mess in there! Dust and crap on everything. The ceiling is falling in places, cracks all over. I'm glad I put those pots under the table."

Rusty grinned. "Good thing you talked me into those eight-foot concrete steel reinforced ceilings, Red."

"Thought they might be useful."

"Good job on getting food ready, Arron. Have the guys started going in?"

"Yeah, Red, the first three were sitting down when I left. Oh crap, sorry, I forgot to bring the rockets."

"Don't worry about it. We'll have plenty here in a little while."

"I need to round up two more guys for the bazookas. Red, who do you think would be best?"

"Best versus available, I'd get Joe and Snake, there's no action on the southside."

"Good. I'll get'em."

"Rusty, get them to eat first."

Rusty gave him a thumbs up.

Another hour went by with very light action. Even the sounds from over at the Ridge were subdued. Booming artillery in the far distance, though, could still be heard. All the men had eaten. Joe and Snake had assembled the two additional bazookas and had them stacked against the wall in fort # 1. Red went over the instructions with both of them again. Red explained what he and Rusty thought would happen next and wanted them ready to move fast.

"I'm taking the outside position over here," pointing to the right of fort # 1, "Snake, I want you at fort # 2, on the left edge and Joe, you got the middle of the trench."

"Great! I've got the best view." Quipped Joe.

147

"You're always getting preferential treatment, Joe. Trips to the city, best views, I'm starting to think that Red's got a thing for you." Snake said to Joe with a smirk.

Joe laughed and shot him his middle finger.

Red was serious.

"Hey, you two! Remember! Prone only, no shoulder stuff. Fire and duck." They both nodded as they went to set up their firing ports.

Foxy was scanning through his sandbag opening with binoculars when Rusty came up behind him.

"Foxy, I need you to check in with your guy at the artillery battery. Try to get an update on what's going on."

He grabbed the walkie-talkie, "Lieutenant Kun, Lieutenant Kun, this is Foxy at the Post." He repeated twice more.

"Foxy, Foxy! My area is now an aid station. I was helping someone. What do you need?"

"Just info, if you have any."

"Bad news and some good news. We're down to two guns, heavy causalities. Captain Hui is still holding the Ridge, and we're expecting reinforcements shortly, but I have no details."

"We're still holding here. Any chance we could get some help when the next attack comes."

"Doubt it. They'll probably attack you and the Ridge at the same time; the Ridge comes first. If things change, I'll let you know." He clicked off. Foxy translated the conversation for Rusty.

"Reinforcements! That's excellent news, Foxy!"

Red came running over, "Rusty, we don't see any troops on the westside, I don't like it."

Foxy looked out his viewport, to the northern tree line. "Nothing here either. Looks like they pulled out."

Rusty let out, "Oh shit! Only one reason they'd do that. We're gonna get hit with their big stuff and it'll be concentrated. Red, tell everybody to get hunkered down. I'll get the guys on the eastside to the rear of the building. Go!"

Red and Rusty took off. Foxy and Tracker left the fort for the trench and got flat.

When Rusty got to the rear, Marty was down in the trench. Rusty started to get prone when Marty asked, "Did you see the men on the road?"

Rusty stood up, surprised. "What men?"

"ROK troops. They were running like hell to the north, to the Ridge most likely. Looked like a lot of guys."

"When?"

"When you started yelling coming down the trench, I looked out onto the road. Just as you got here, I saw a jeep go by, you know, the one with the cannon."

Rusty looked at Marty with a disbelieving expression. Then said, as he turned and looked out toward the road, "You're not joking, right?"

"Jesus Rusty! I wouldn't joke about that!"

"I don't see anything, no movement."

"Well, they weren't marching, they were running."

The shell detonated fifteen feet in front of Rusty, throwing him to the trench floor with a splash. He felt like he was hit with a baseball bat.

The constant rain had started to fill the trenches. Water was now about an inch above the wood plank flooring. Mud from the trench walls began accumulating in with the water on the wood flooring, where men now had their bodies pressed tightly, as shells started bursting all around the Post, on the building and then everywhere.

Rusty became disoriented and started vomiting into the mud. Choking on the muck, he instinctively turned to breathe, then covered his ears with his hands from the noise of the hellish explosions, then curled up in a fetal position. He felt sanity slipping away, with no idea how long that incredible noise kept up.

Then it stopped.

He lay there a while and slowly became aware. Somebody was shaking him, yelling something. Opening his eyes, he felt the rain hit his face and then saw Marty leaning over him.

"Rusty! Can you hear me?" He could vaguely make out the words.

Marty did a quick exam to determine if anything was broken, then helped him up into a leaning position against the trench wall, checked for any wounds, gave him a canteen and held it to his mouth. He took a swallow.

He felt nauseated, but his mind started to clear. He finally spoke.

"Not sure I ever want to experience that again. Am I wounded?"

"Not that I can tell. No external blood and no coughing up blood. All good signs of no internal injury."

He looked around, noticed men getting up, walking. Big chunks of concrete lay nearby.

"Okay, I'm feeling better. Help me stand up."

On his feet now. He felt wobbly.

"Any causalities?"

"Don't know. I think someone's hurt over at # 3 fort. I wanted to get you squared away first. You okay if I leave you?"

"Yeah, yeah, go, I'm okay."

Red came over to Rusty from the fort # 3 site. "You look like shit. You all right?"

"Shut up. Feel like shit too. How'd we do?"

"Not so good. We took a direct hit in the trench on the westside just north of # 3, killed Ray. A concrete slab came off the building on the north side and killed another Ranger, Jeff Martin. He wasn't supposed to be there but wanted to talk with Ben. Ben's a little freaked out about it. That's two friends Ben's lost today. A few guys are still groggy, but they'll be all right."

"Three dead so far, damn it! Ray's dead? Shit!" He moaned, "Oh, my head hurts like hell."

"You'd better stay still."

Then he remembered. "Red, Marty said something to me right before the shelling, about him seeing ROK troops on the road going north. Did you see anything?"

"No, but if he saw them, that's great news. I just hope they got settled before the shelling."

"You better get ready for the next curtain to rise. It shouldn't be too long." He leaned back against the trench wall. "Think I'll stay here a while."

"Right. Stay put!"

Korea - The Post – 1130 – (11:30 am)

"Troop movement in the trees, both sides," Arron yelled out. Red had just finished fixing his bazooka firing position. The men had busied themselves repairing the extensive sandbag damage from the shelling all along the whole trench perimeter and had just finished.

"What is this, Red? They're laying smoke along the entire front, including the open field."

Machine gun fire opened up from the tree line on both sides.

Red went to Tully in the mortar pit.

"Go check Rusty, tell him what's going on. I think we should bring up most of the men from the rear. They're concentrating their attack across these two fronts. Not sure Rusty should move. He looked messed up."

"Okay." He started to move, then turned back. "Red! I've never heard of the Russians using smoke before."

"Must be something they picked up from us."

"Yeah, maybe. I'll go check on Rusty."

He listened to Tully, nodded his head, but didn't say anything; otherwise, he didn't move. Tully made the decision and ordered every man, but three, up to the action zones. Moving quickly with the men, he worried about Rusty, thought maybe he was just shell shocked, but still, maybe it was worse. So, when Marty came out of the mortar pit, Tully grabbed him. "Listen, Rusty is in bad shape; you need to go back to Rusty and help him."

"Can't now, I've got to defend this position, but I'll go just as soon as this threat is dealt with."

Smoke shells were coming in a serious barrage, but the rain dissipated the lingering effect, so they had to keep up a massive bombardment to maintain the effectiveness of the smoke.

Above the sound of the machine guns, Red heard dull explosions. The mines, he thought, they're trying to remove the mines behind the smokescreen.

Then he heard another sound, one all too familiar. The clanking of tank tread, unmistakable. The worst sound an infantry soldier can hear. Son of a bitch, he thought, I'm ready for you, but still, Jesus, it scares the shit out of me.

"Tanks!" Red yelled, "Everybody! Stay down!"

"Bazookas! I'll let you know when they clear the smoke."

The sound of the tanks firing began.

Crazy, Red thought. They can't see anything; the tank gunner is just guessing where to fire, think they first want to get very close to help clear the mines for the infantry. Good luck. I'll just wait for you.

"Red!" Tully yelled above the noise, "any idea how many?"

"Somewhere between two and ten." He yelled back.

"Big help!"

"You think they can keep up this smokescreen? I've never seen so much concentrated smoke, in the rain no less."

"Me neither. Fifteen minutes so far. The tanks got to be near the second wire. From the number of mines going off, I'd say the tanks are doing a good job of clearing them."

To the left of fort # 2, Dean and Ben were near each other, squatting in the trench.

"You shouldn't be near me, Dean; I'm bad luck."

"Don't be an ass, Ben! Sam and Jeff were my friends too. We've all been together for a long time. We knew this could be a shit mission. Ain't nobody's fault."

"Those tanks can't hit crap, can they?" Dean said, wanting to change the subject.

"Yeah, sure glad too. Kind of reminds me of Hill 400 at Castle Hill, remember? Except now we're on the hill. I think we're better fortified than the Jerry's were. They held us off for four God awful days!"

"You're right. Does remind me of that fight. But that hill was much larger, maybe a hundred feet high, this is what, thirty feet?"

"How many tanks did we lose? Ten, fifteen? I don't remember, but a lot, right? Our tank's fire was ineffective, shooting up the hill, they couldn't kill those 88's. Just like here. Just a bad angle of fire."

"Let me remind you that we did finally take that hill."

"Right! Rangers took it. I don't think there are any Rangers out there."

Hard to beat that logic, Dean thought.

A few shells from the tanks hit the building, but most either missed everything or hit the hill in front of their positions. The noise, however, was continuous.

Kaboom!

"That was one of my special mines! I think they're inside the second wire." Red offhandedly said to Arron.

"How many kinds you got out there, Red?"

"I got four types of anti-tank mines, Arron. Don't you remember? You laid a lot of them."

"Nah, you see one you've seen them all, I don't remember."

Kaboom!

"Those are my favorites, German. Big charge. Took us a while to figure out that you can't remove them by hand. They packed a secondary charge if you tried to lift them out."

"Bastards" was all Arron could say. Then said, "How'd you get them."

"We found warehouses full of them."

"The Brits made a nice one too. Got a bunch of those. We made two kinds I liked, and they were plentiful."

Tully came over to them, "Red, I think the tanks have a lot of troops behind them. How about I dump some mortar rounds on them."

"You got enough ammo?"

"Oh yeah! Colonel Nelson gave us a lot!"

"Then I say go. I think they're going to run out of smoke shells soon and it's going to clear. Then the real fun begins. Go start the party early." Tully headed for the mortar pit.

"Set your distance for four-hundred yards, aim for the open field between the trees," he pointed from memory.

A few minutes later, thump, thump, thump....

Then the smoke shells stopped falling, and the field started to come into view.

Red yelled. "It's clearing! Get ready!"

Enemy mortar shells started hitting around them.

Dean and Ben moved into their firing positions. Dean set his rain tent, looked into his scope, and said to Ben, "Time to go to work."

Bang

The tanks became visible first. Dean's shot had taken out the commander now slumped over the turret of the closest easterly tank. He could see six tanks, spread evenly in attack formation, all firing. He couldn't see behind them yet.

"Bazooka teams, fire at will!" Red yelled, then moved quickly into his spot, glanced down the trench line to see all the men adjusting to their firing ports. He lifted the weapon into position, aimed at the most westerly tank and fired. The tank exploded. Saw part of the guy in the turret fly in the air. He hunkered back down, then manhandled another rocket into the tube, set the ignition sequence, and repositioned the bazooka back into the firing port.

Ben was firing his rifle at the men flat on the ground at his front, which were still trying to clear mines between the first and second wire. Off to his left, the lead tank in the middle of the attack column had made the most progress, around ten yards inside the second wire, where it had hit a mine. It lay still with one of its tracks off but was continuing to fire.

Joe and Snake fired almost simultaneously. Hard to miss at three hundred yards. They didn't, then ducked down and started reloading.

Red fired his second round. Four dead tanks, two to go.

Foxy wanted to remount his machine gun but decided it wasn't time yet. Grabbing his rifle, he found a slot and started firing.

Dean moved to a spot between Joe and Snake and set up. He saw lots of fire coming from the woods, he swung his rifle over, focused on that area. Sure enough, he thought, my first sniper. Bang. He kept finding targets. Bang. Bang.

Tully clearly saw the men behind the tanks now, but not many. His mortar team had thinned their ranks pretty well, he thought. He saw bodies lying in the field to the rear.

"Stop firing! Manny. Shift the mortar to our front along the tree line. They're massing again."

All six tanks were now burning.

"Red! Let's set the machine-gun!" Yelled Arron.

"No! Use your rifle. We can't use it yet; we're too exposed." Arron understood. They both grabbed their rifles and aimed at the men around the destroyed tanks.

Mortar explosions, heavy rifle, and machine-gun fire were continuous around both trench lines. The fighting lasted another ten minutes or so then, slowed, finally stopped. The few remaining soldiers left outside of the trees withdrew.

Ben had moved over with Dean and set up near Joe. Both were looking over the battlefield. It was a horrifying, awesome sight.

"Joe, you fought the Japs, right? You ever see mass killing like this? In one small battle, in what, a few hours? Because I never saw anything like this in Europe."

Joe scanned the burning tanks, maybe two hundred bodies all around them, then over in the field by the woods, where many more bodies lay.

Joe lowered his head. "Yeah, Ben, I have. They were our boys. All dead in less than five minutes. Okinawa was a ruthless place." Joe shook his head, trying to shake the memory away.

"Jesus, Joe, I'm sorry."

"Medic, medic!" The call came from Moose, in the western trench. Marty came running. Moose held a large compression gauze over the side of Mario's face and ear.

Marty pushed Moose away, knelt, pulled off the gauze, and started to clear away the blood so he could see the wound. Mario was limp, unconscious. The rain helped him clean up the bloodied area, and saw that it was a bullet crease that was only skin deep, running from the corner of his eye to the back of his skull, six inches. He breathed a sigh of relief.

"He's okay, Moose, just got a bullet crease on the side of his head. Pretty damn lucky, I'd say. Come here, lean over, make a covering over his head so I can dry it and put on a clean dressing. He'll have a hell of a headache for a while. Get another slicker; try to keep him dry as possible. He should stay lying down, might have a concussion."

"Easier said than done, Marty. We're all soaked to the bone." Moose lamented.

Chapter Twenty Five

Sunday - June 25th, 1950   Day of the Invasion
Japan - Eighth Army Headquarters – General Walker's
Conference Room – 1300 - (1:00 pm)

Present in the room:
General Almond – General MacArthur's Chief of Staff
General Walker – Eighth Army Commander
General Dean – 24th Infantry Division Commander
Colonel Harris – Eighth Army Executive Officer
Colonel Nelson – Eighth Army G2 (Intelligence Officer)

General Almond stood, facing the seated officers.

"I requested this special meeting in light of current events in Korea. In the past few hours, our staff has been told tall tales of intrigue and conspiracies, of falsehoods wrapped in half-truths and most troubling of all, few facts. It seems, Colonel Nelson, that you're the only credible source of real information. Your earlier reports of aggression by North Korea are proving to be most accurate. General MacArthur needs to brief the Joint Chiefs and our civilian leadership on these events and put forth options for our response. So, Colonel, your input, as well as from those in this room, is critical in determining what happens next. You're known for being a straight shooter, and somewhat of a maverick. You see things before most others do, and those assessments have turned out to be more often right than wrong. Your honest evaluation of these events and the foreseeable future potential is most needed now. It will not fall on deaf ears."

General Almond moved toward a seat to the right of General Walker.

"Please proceed with the briefing, Colonel Nelson."

Colonel Nelson was a bit taken aback by the General's comments, particularly the deaf ear part. His experience with general officers was that they rarely, if ever, admitted to any failings. He walked to the map on the front wall.

"As of noon today, the general invasion by North Korea has made substantial progress. The seven North Korean divisions we had identified a week ago are all engaged in ongoing battle across the entire North-South border. Here on the east coast", raising his pointer to the large map on the front wall, "an eighth division staged an amphibious landing at Yangyang," indicating the exact spot with his pointer tip, "ten miles south of the border; cutting the coast highway and engaging the rear elements of the 8th ROK Division, essentially trapping this division."

"Here, on the other coast, equal success was achieved by the North. In a two-pronged assault, they pushed the 17th ROK Regiment into the Ongjin Peninsula, trapping them. The second prong crossed the Yasong River, decimated another ROK Regiment, and captured Kaesong five miles south of the border and a key access road down the west coast, with a straight shot to Seoul, forty miles away. However, they still have to cross the Bukhan River, south of Kaesong. This river merges with the northern section of the Han River, and its the last defensive position the ROKs have to impede the capture of the capital."

"Colonel, can the ROK troops hold this defensive line at the Bukhan River?"

"No, General Almond. I believe it will merely be a delaying action, the length of which, I hazard to guess, will be short. Maybe a day, two days at best."

"Are you saying that the South Korean troops are inferior and not capable of fighting?"

Colonel Nelson noticed a grimace from General Walker at General Almond's question, hmm, he thought, I've got to find out what that meant.

"Well, in a way, yes. However, I don't think it's in the way you mean. I think the ROK soldier is as capable as any soldier in the world; they just have not had the training required to become so. Neither has their leadership been educated adequately. Of course, there are exceptions which prove my point. I'll get to that in a moment."

"Are you saying that Brigadier General Roberts has not performed his job satisfactorily?"

"I'm a Colonel, General; I don't judge generals. However, I will compare resources applied to both the North and South's military training programs; you be the judge. Brigadier General Roberts, as head of KMAC (Korea Military Assistance Command), was given less than five hundred men to train the entire ROK Army. Only two hundred were actual trainers, and most can't speak Korean. The Russians, on the other hand, provided five hundred men as well, but for **each** of the North's ten divisions. I could go on about the tanks and artillery the Russians supplied, well, I could go on for quite a while, I guess."

"So, the reports concerning the readiness of the South Korean Army are not accurate?"

"I recall a recent report, three days ago I believe, that KMAC issued. Stating that, and I quote, *the ROK Army is so well prepared that should the North invade, it would be like taking target practice.* I suggest, General, that the answer to your question lies in the question."

General Walker thought, well-done Colonel, well done.

"We do have positive news in the Uijongbu Corridor." Colonel Nelson moved his pointer to the center of the map. "The ROK forces here have done a magnificent job holding up what we think is the main attack force on Seoul. Good leadership and careful positioning of troops have denied access to the vast corridor at a critical chokepoint three miles south of the border. Farther east,

159

about ten miles, near the town of Cheorwon, a strong defensive position has held back the initial assault. But the area lends itself to the use of armor, which has not been employed as yet, but I suspect that it will be soon. When that happens, the defense will collapse."

"Have we seen much armor engaged? They can't have that many tanks to make a difference?"

"So far, reports indicate sporadic use of tanks, but when involved, they've overwhelmed the ROK forces, because they have no weapons that can stop them. We don't know the exact count of tanks, but from intercepted messages, spies, and deserters, we have a pretty accurate picture of their Order of Battle and armor strength. We believe they have positioned upwards of one hundred and fifty T-34 tanks for their main thrust on Seoul."

"That doesn't sound right, Colonel. I'm not aware that the Russians supplied that many tanks to them, are you sure about this?"

"General Almond, I'm confident about the intelligence. We have multiple sources as confirmation. I believe I furnished your G2 with these reports many months ago when we ascertained that the Russians had shipped in over two-hundred-fifty T-34s in the past two years. That was part of my request to provide the ROK Army with heavy defensive weapons, like tank destroyers, to offset this threat."

General Walker kept a straight face but thought, be careful here, Colonel, nobody likes the blame game, and this thing has just started, they'll be plenty to go around before this is over.

"I don't recall that report. I do wonder about the detail of your briefing, however. How is it that you know so much about what's going on?"

"If you must know my secrets, General, you must honor my request for discretion in sharing this information."

General Almond smiled at this. "What? Honor among thieves? Okay, I agree to your request."

"Thank you. I didn't go through regular channels, General. I established a direct relationship with my counterpart in the South

Korean Army, Colonel Pyong Oh, a competent officer. We have communicated almost every day for many months, but for the last two days, it's been more like every few hours. He established a strong staff and field officers that are attached to most of the front line units. He has broad powers and can direct units to respond to emergencies without going through his headquarters bureaucracy. He's been at his desk for the last 36 hours. I spoke with him thirty minutes ago."

"So, what is happening in the capital?"

"Panic, in his words. The politicians are leaving the city as fast as they can." Momentary chuckles from the officers.

"Of course, he is redeploying his forces to meet the immediate threat against Seoul, but he and the rest of the Army Staff are realistic, this is only a delaying tactic. He concedes that the capital will fall within a day or two, hopefully, longer, but the Army's focus is on establishing a strong defensive position south of the Han River. They'll defend the capital to slow them down, but if the North Korean's can cross the Han, the country could collapse."

"I see. Not a very optimistic outlook, probably very much in panic mode, something to stir up America, I think. I recently found out that you have some observation Posts established near the border, have you heard from them?"

"No, sir, I haven't had any word from them since the action commenced."

"Thank you for your briefing, Colonel."

General Almond turned in his seat towards General Walker.

"General Walker, what's the status of the Eighth Army, should you be called upon to take action?"

"As you know, I placed all four divisions in Japan on high alert three days ago, so we are ready to respond quickly. I've asked General Dean to sit in on this meeting since his division would be the first to go. If you have any specific questions for him or other questions concerning the rest of the Eighth Army's capability, Colonel Harris will answer those."

"Thank you General, I would like to know the combat strength of your division General Dean, and how much armor is attached to you?"

General Dean stood from his chair, moved forward three feet, and turned to address General Almond.

"The 24th Division roster shows 7,846 combat-ready troops in three regiments, plus we have a Light Artillery Battalion with a total of twenty-four 105mm guns, and one Medium Tank Company consisting of seven Sherman 75s."

General Dean felt dismissed and returned to his seat.

"Colonel Harris, what's the status of the rest of your divisions?"

"General, the remaining three divisions in Japan are similarly constituted, but less so in all aspects. As a guide to the total combat strength of the Eighth Army in Japan, think in terms of twenty-five thousand men with no heavy weapons, few tanks and not much in the way of ammunition reserves."

"Sorry, I asked the question." He said to no one in particular. He then turned to Colonel Nelson.

"To wrap this up, Colonel, I want your assessment on the ability of South Korea to stop this invasion."

A little shaken by this question, he stood up.

"No offense, General, but I thought I made myself clear. The South Korean Army is being overrun and is on the brink of total collapse. The capital will fall in a few days no matter what we do now. But if we don't do anything very soon, the whole country will be overrun in a matter of weeks or even days if they see no hope. They cannot survive without substantial assistance."

Chapter Twenty Six

Sunday - June 25th, 1950   Day of the Invasion
Korea - The Post – 1430 – (2:30 pm)

Rusty sat in the rear trench through the last assault. When it ended, he made his way to fort #1, where he met Red and Arron, leaning against the fort wall.

Red was taken aback at seeing Rusty wobble over, his face ashen.

"You need to rest, buddy. Why don't you go into the building and get some shuteye? Mario's in there doing the same thing, he's by the door in case the shelling starts again. Go with Arron."

"Good idea." He shook his head a little. "I'm still not a hundred percent from that shelling. Don't feel right." He didn't move.

"Marty told me he thought you might have a mild concussion. You sure look like you do."

"Okay, I'll go in." Rusty uneasily started slowly walking down the trench.

Arron got up and started following. Seeing him wobble, he decided he should walk next to him.

"Hey, Rusty, wait up." Rusty stopped and leaned back against the trench wall, looking disoriented. "You okay?" Then Rusty collapsed, but Arron grabbed him and started carrying him down the trench. He yelled down to Larry to give him a hand; they both got him into the building.

"Get Marty!" Larry took off.

"He just passed out," Arron told Marty.

"I thought he might have a concussion. He needs to rest. Let's get him comfortable." He saw Mario laying a few feet away, his eyes open, looking at them. "You're awake! How'd you feel?"

He slowly responded, "Not sure yet, the room is spinning, and my head has a church bell in it. What happened?"

Marty smiled at him, "You sound fine. You know you're a very lucky guy. Another half-inch, you'd be dead." He told him what happened.

Tully came over to Red. "Nothing for the last hour, why don't we get Arron to heat some food, maybe coffee? The guys could use something hot. Get them out of the rain for a break."

Red nodded, "Could use a cup myself." He said solemnly. He was deeply worried about Rusty.

Tully found Arron, told him what he wanted, and left to resume lookout.

Arron went back into the building, Rusty and Mario were lying on the floor to his left as he started for the mess room passing Rusty's office and paused, what a bunch of crap, he thought, big chunks of concrete on the floor, big cracks in the ceiling, what a shame. He shook as he glanced at the three covered bodies in the corner. Damn! I liked Ray.

He went to the mess room, started setting up. He kept himself busy, reheating a big stew pot, making coffee, putting another pot on so maybe the men could take a shave. He kept moving.

His mind wandered as he was doing all this. I really like these Rangers, they all seem like good souls. They've been through stuff in the big war nobody would believe, survived, and still managed to keep their humanity, and they still care. He suddenly stopped, realizing he wasn't just describing the Rangers.

A few minutes later, Snake came into the building to check on Mario.

"Hey! You're up! I was worried. Marty said you'd be okay, that you just got a new tattoo on your head."

"I hope it compliments my good looks because it's not doing anything for my disposition."

"You're definitely feeling better. Hey, how's Rusty? Have you talked to him?"

"He's been mostly out, stirs a bit and moans. Marty thinks it's just a bad concussion, but I'm worried."

"Best thing for him is to rest, you too! I gotta get back. See you later."

"I can't raise the ROK Artillery unit. Not sure why." Foxy said to Red and Tully.

Red was determined."Keep trying; I'd sure like to know about those reinforcements."

Dean had kept busy through the semi-lull in the fighting. Quite a few NKs had not retreated, or couldn't and had instead taken cover in the many shell craters in the field. Trying to keep from being bored, he focused on these men. He'd watch a hole, seeing an occasional head pop up, actually studying their behavior. Curious, he thought. If I were out there, I wouldn't move. Why do they want to look at us? It's not like we're going to come after them. And then he would see a guy raise his rifle over the side of the crater. Why? Bang.

He decided not to shoot anyone that was not aggressive. It didn't slow him down any. His exception was officers. If he could spot them, they were always the number one target, right along with snipers.

"You enjoy your work, don't you?" said Joe, sitting in the trench next to him.

Dean didn't move from his scope but responded, "My job is to save my men. The less of them, the better our chances are."

"You didn't answer my question."

"Why don't you be honest with the question?"

"Okay. Do you like killing?"

Dean moved from his position and sat down.

"Joe, let me ask you a question first."

"Shoot. No pun intended."

"Ever shoot a man in the back?"

"Of course I have. Lots."

165

"Then what's the difference between you and me? Because I think you think I'm an assassin with no soul."

"No, it's not that. I mean, I've killed in battle—men shooting at me, trying to kill me. But you just lay there, take careful aim, and coolly kill. They don't even know who shot them. That's weird to me."

"So you think killing has a moral aspect to it. Like it's okay to kill when the enemy's engaged with you, but it's not okay when they're just "planning" to kill you?"

Joe was quiet for a moment, finally said, "Killing is killing, isn't it? I guess it's all good as long as it's the enemy."

"You wouldn't make a good sniper."

"I know. I have other redeeming qualities."

They both grunted.

"I'm curious, Dean. You've been shooting for the last few hours, what, maybe thirty shots. How many did you score?"

"It's thirty-two shots, Joe, and I don't miss."

Korea- The Post – 1800 – (6:00 pm)

Rusty opened his eyes. He felt foggy. Where am I? What happened? I was standing by the trench wall. Oh, my God, my head hurts! Don't raise your head again, big mistake. Oh shit, I'm spinning.

"You awake, Rusty? It's Mario, right next to you."

He couldn't talk. Who was Mario? He passed out.

Tully had made the rounds three times since the last action ceased. Each time, he spent a few minutes with each man. So far, everybody was doing okay, considering. They were all interested in the "next" thing to happen. I wish I had a clue, he thought. He decided to spend a little time with Red, so he went over to fort # 1.

"Hey, Red. Been thinking about our machine guns. How about we set'em up on the fixed poles, get them ready to fire. Maybe we're in for a tough night."

166

"Yeah, been thinking about the same thing. Okay, I'll get'em ready when it gets dark. Have you checked in on Rusty?"

"Three times, and spoke with Marty the last time. Said he's checked his head and couldn't detect any swelling. A good sign, he's sleeping, and that's what he needs. Mario's awake; his brain is still ringing, but he's good. Lucky."

"Yeah, lucky. I'll tell you what's lucky it's you showing up! You changed everything. We wouldn't be chatting here if you hadn't arrived, that's for sure. I thought you were pretty sane before, you know. Now I'm not so sure, volunteering for a suicide mission. Don't get me wrong; I'm really glad you're here, but ...... this is bad, and you knew it would be bad. Why did you volunteer?"

"Shut up, Red. By definition, we Rangers are insane. I couldn't stand being in Japan listening to how much fun you were having over here, so don't ask again."

"Jesus! Don't be so touchy!"

Tully was upset he went off on Red, decided to change the subject.

"I want to tell you what a magnificent job you've done here, this is an inspired defensive position. The German High command would be proud of your accomplishment."

"You noticed."

"Bad memories aside, you had to admire their brilliance."

The gloomy daylight gave way to a very dark and rainy night. The heavy shelling started up once again.

Korea- The Post – 2330 – (11:30 pm)

The shelling had slowed to intermittent detonations a few hours earlier, then stopped altogether.

The rain stopped about an hour ago.

All quiet.

Too quiet, Tully thought. He couldn't see anything as the night was made black by the lingering rain clouds. He went to Dean. "Can you see anything?"

Dean adjusted his rifle and scanned the front, side to side. "Can't see shit. Too dark."

Tully left for the mortar pit, rounded fort # 1, and stopped. "Hey, Red, I don't like this. Too quiet. I'm going to send up a flare."

He yelled down the trench line. "Flare going up, get ready."

At the end of the trench, at fort #3, "Snake, what did he yell?"

"I just heard *ready*. Get on the gun, Larry!" He jumped up and swung the machine gun into position. Fort #3 was all set.

Thump. The illumination shell opened at about four hundred feet high and five hundred yards out, igniting a 145,000 candlepower flare attached to a parachute. It slowly drifted down above the open field between the woods. It lit up the entire western and northern fronts.

On the north side, Dean took the first shot. An officer was standing in the field at his front, less than three hundred yards away, directing the men ahead of him, as they were crawling, and clearing mines. Bang.

Foxy opened with his machine gun at fort # 2. Then everybody joined the madness. Heavy firing erupted from both enemy treelines. A second parachute flare went off.

Arron thought he was back on the farm, shooting rats in the field. The prone NK soldiers didn't know what to do, stay low, get up and charge, retreat, or keep clearing mines. Their officers were yelling at them. They started dying fast. Bang. Some men got up to run. Bang. Some decided to stay and fight. Bang. Others buried their faces in the mud. Bang.

"Oh, brother!" As the first flare lit, Larry saw a cluster of soldiers kneeling just behind the second wire along the western front, kind of like deer in the sudden light, as he swiveled the 30 cal machine-gun into their ranks and opened fire. Snake handled the ammo belts as the pro he was, keeping up with the furious firing pace, anticipating changing the belts in seconds; it was exhilarating, but he desperately wanted to know what Larry saw. He'd find out later, he thought, if they got through it, so he just kept feeding the ammo belts.

In the chaos, some men started to charge the defenses, straight into the minefield. Boom. Boom. As the parachute flare drifted down and extinguished, another flare ignited.

Joe kept firing. The M-1 rifle was a fantastic weapon for this purpose, he thought. Just like at a target range. At two hundred and fifty yards, 80% accuracy for the average shooter. I'm not average. Every three seconds. Bang.

It ended as suddenly as it had started, the attack was over.

Chapter Twenty Seven

Monday - June 26, 1950, Day Two of the Invasion
Korea - Somewhere South of Kaesong on the Western Invasion
Route – 0030 – (12:30 am)

Sergeant Lopez huddled in a dense cluster of bushes on a small ridge near the northern edge of the Han River Junction. He talked to himself. What's it called? He tried to remember his trip with Stan when they scouted out their withdrawal. The Bukhan River, that's it! It meets the Han River west of here, then flows to the sea. There was a bridge nearby, further south, I think.

Flashes of shell bursts lit the far side of the river, but he couldn't see too far south of his position. The bridge is probably blown by now. Damn it! I wish I knew exactly where I am. Parts of this river are pretty narrow. But man! With all this rain! This river is going to be moving fast. Not sure I can swim across that kind of current.

He rested, took inventory of his day. Damn lucky so far. Twenty-four hours, and I've only gone twenty miles. Shit. Well, he said to himself, you wanted to avoid all contact, and then this terrain. Jesus! It's bad enough trying to make your way on foot during the day. Last night and in the rain, it's a wonder I didn't break my neck. Then the shelling all day. South Korean troops running all over, then finally, the enemy showed up. Goddamn Commies are moving so fast. Now they're dug in all along this river. I can't get in front of them now. No way am I doing any swimming at night. Another good reason to sit tight. Are those

ROKs over there? They've got to be scared shitless. They'll shoot at anything that moves. Okay, George, time to take a rest. Just wait till something good happens. He fell asleep.

Korea - The Post – 0100 – (1:00 am)

"I want you to replenish all the M-1 bandoliers, and Marty don't give me any shit about it!

"Anything for you, Red."

"We're just about done with the mortar ammo, maybe ten or so rounds left." Tully sat down next to Red, both exhausted.

"I just saw Rusty. He's awake and lucid. Says he feels better. I told him to rest more."

"I took it upon myself to tell the men to try and get some sleep, leaving one man in each trench to stay on watch for one-hour shifts. Hope you're okay with that, Red."

"Cut the shit, Tully! You make solid decisions. I got no ego problems. As far as I'm concerned, we three make a great team."

"Any luck with raising anyone?"

"No. Foxy has been trying every hour. We need to find out what's happening pretty soon. Let's go find Larry; maybe he's got an idea."

They found Larry in the building under a table working on something by flashlight. They wanted an update on getting communications fixed.

Dismayed, Larry said, "We've got some problems." Sticking his head out from under the table, he looked up at them. "Our hardline cable connection is broken somewhere between here and the road. I know it's not in the road because that cable is buried ten feet deep. I've spotted two likely shell craters where the line could've been cut from there to here. One's about twenty yards down toward the road; the others about ten yards farther. Sure glad we didn't lay it through the minefield."

"Can it be repaired? How quick?"

171

"Yeah, Tully! It can be repaired. I got extra cable. How fast is another matter. If it's a clean break and we don't have to dig it out, then it could be spliced together in say, half an hour or so. But it could be cut twice, or more. But I'm not finished with my problems. The line is one, but this console I'm working on is another issue. It got knocked off the table by some ceiling concrete and got damaged. That's what I'm working on now."

"Do we have a town hall meeting going on under the table?" Rusty stood in the hall doorway with Mario giving him support.

Tully spoke first. "Only healthy town folk are allowed! You guys should be resting."

Responding, Mario made light. "I'm never coming to this Hospital again, way too noisy, no pretty nurses, men hiding under tables. This is a weird place."

"What the hell are you guys doing?" Rusty inched along the wall and then sat on the floor across from them.

"Trying to fix our communications." Tully looked back at Larry, "Go on. What's wrong with the console?"

"Looks like one of the tubes got damaged. I was just about to replace it. I'm hoping that's it."

"When will you know?"

"Right about now." He flipped a switch, and the console lights came on.

"Okay! One problem solved. What about the VLF. Can we get that to work?" Asked Red.

"I don't know. The turret got destroyed. The dish unit disappeared in pieces. I have a mini portable dish, but it's the only one we have. Don't want to put it up on the building, too dangerous. We'll need it for later…. If we get out of here."

"Cut that *IF* shit! We'll make it out!" Red snapped.

"I'd like to try repairing the cable," said Larry. "Red! What do you think?"

"I think it's near suicide, but I don't see any other option. Rusty! Got any thoughts on this?"

Rusty raised his head. "Right now, I'm just happy I know everybody's name."

"Go lay down, Rusty. Mario, how're you feeling?"

"Peachy. Heads clear, it just hurts. Not wearing a steel helmet for a while, though. I'm ready to get back."

"I'll get Marty. You can return if he says so. Otherwise, you stay with Rusty."

Mario sat down next to Rusty, who was stretched out flat on the floor.

Larry spoke up. "I'll need Joe with me. He's familiar with splicing lines. Maybe two more guys to help dig or hold a blanket over us to hide the light. You guys pick them. I'd like to go before the next shelling begins. You got that schedule?"

Red fumbled with his shirt pocket, "I know I have it somewhere? Smart-ass!"

"I think Snake and Arron should go." He said.

"Larry. Get your stuff ready! Tully, get the rest of the men and organize this. Set out ASAP. I think the NKs are done until they figure out their next move. Let's take advantage of this lull."

Korea - The Post – 0145 – (1:45 am)

The men belly crawled to the first crater. Dense clouds hid any potential light from the night sky, but occasional shell bursts nearby cast dim quick flashes. Even so, they were protected from the enemy lines to the north and west. The real danger for them was an artillery shell.

Larry crawled into the shell hole, then signaled for the rest to follow. "Snake, Arron! Put that blanket over us!" Larry turned on his flashlight with his hand over the lens, then opened his hand slightly to let out a small ray of light.

"Not too bad a break. Give me a hand clearing the ends, Joe. Guys, get settled up there. We're gonna be a while."

Repairing the damage, they finished in fifteen minutes and backfilled as much loose dirt as possible to cover the exposed cable.

"You guys stay here. I'll check the other crater." Larry crawled off. He was back in a few minutes.

"In luck. No break there. Let's get back."

Back in the building, Larry established contact with Colonel Nelson. After being briefed about the general state of the NKs progress, Red was finishing his update of their situation. Tully was listening in on the conversation.

"Those new bazookas are fantastic, Colonel! Stopped those tanks cold. Send more! We gave two to the ROK troops and I think that's the reason they're still holding on at the Ridge."

"Good to hear! I'll see what I can do. Is Rusty okay?"

"Like I said, Colonel, he's pretty groggy but Marty is watching him like a mother bear. Says he should recover with some rest."

Red listened to Colonel Nelson's response.

"Right! Colonel, I'm not sure what their next move might be, but I'm sure it'll be interesting. My primary concern is that the ROK position on the Ridge will be overrun. That would expose us to a three-sided attack and probable encirclement. If the ROK can hold, we could prolong this action for another day or two."

He paused, Nelson responded.

"Yes sir, I know. We'll do our best." He clicked off.

"Interesting. We're the only area that's held up the invasion. Captain Jordan's old area around Cheorwon folded when tanks showed up a few hours ago and ROK command lost contact with their units defending the town. He wants us to hang on here. Every day we can hold them off is significant."

Chapter Twenty Eight

At least something's working right. Nelson was thinking about the Super Bazookas.

"Sergeant! Get me, Colonel Harris! It's important."

"Line two, sir! Colonel Harris and he's grumpy."

"Sorry to bother you at this hour, Colonel, but something came up that I need to talk to you about."

"What's going on?"

"Those Super Bazookas we sent to the Post have been extremely effective and probably why we've held up the attack in the central valley. I know you have more of them. Can we get them to the ROK troops?"

"Yeah! I remember. It's the only new weapon we have." He coughed a few times. "Well, we didn't get many, and I'm sure glad we got your guys a few, but I don't have anymore."

"What do you mean? You had ten or so and were getting a big shipment in! Where are they?"

"Well, we gave two to the Rangers, if you remember. As for the rest? General Almond thought Chiang Kai-shek might need additional weapons to fight off those water crossing tanks the Chinese Commies might throw at him, so he had them sent to Formosa yesterday. The big shipment is still on a boat somewhere."

"Son of a bitch! Is Almond insane?"

"I don't know about his mental health, but his judgment sure stinks."

"You said more were coming, do you know when they'll be here?"

"Not a clue. I only know it's on a regular cargo ship out of San Diego."

"Would you find out if we could get any flown in? We need this weapon."

"Right. I'm up now, might as well wake a few more people. I'll see what I can muster up, let you know."

Korea - The Post – At the Same Time – 0230 – (2:30 am)

Arron suddenly burst through the entranceway of the building. "Red, Tully, you've gotta come to the fort! Something's going on!"

They rushed with Arron to fort # 1.

"What is it?"

"Listen!" They got quiet. Then heard a distant rumbling sound.

Red focused on the noise. "Not a tank sound. More like construction equipment. Maybe trucks are dragging stuff."

"What the hell is that Tully? It sounds familiar?" Tully strained to listen better.

"I'm not sure. Whatever it is, they're moving across the open field to the west. My guess is they're going to that area behind the tree line."

"Yeah, I think you're right."

"Ah, I think I got it." Said Red. "Remember at the firing range, at that Weapons Evaluation program we had in Japan when we waited for the heavy weapons to be towed out? One group stands out to me that sounds just like this. Remember! The Russian tracked vehicle towing the 82mm mortar."

"That's it! I remember that sound. It came across that field, about a mile away from us."

"I also remember what a powerful weapon it was. What were the specs? I think it fired a six-pound shell with a range of something like three thousand yards."

"That's right! It required a four-man crew! Yes, I remember it clearly."

"We rated it as the best heavy mortar in the entire War. Shit, Tully! We're screwed! From the sounds, I think they're bringing in a whole company. How many would that be, do you remember how the unit would be constituted?"

"Vaguely. I think it was three or four mortars per platoon and three or four platoons per company."

"So, best case, they have nine mortar teams. Worse case, they have sixteen. Damn! Either way, they'll standoff and blast the shit out of us. Tully, I think it may be time to consider withdrawing before we get buried."

"I agree, we can't sit here and let them pound us. Those mortars are accurate, not like the big artillery they've been using. At this range, we're dead meat."

Tully was in deep thought, weighing the options. He became still, focused on something he remembered hearing just recently. He smiled and said, "I have an idea. Let's go see Rusty. You come too, Arron. You're our salvation."

Arron was shocked. "I am?"

Red kicked Joe's boots, "Hey Joe! Wake up!"

"Goddamn it!" Joe said, startled awake. "Rude, just rude!" He got up from the trench floor.

"Take guard duty at the fort, Joe; we'll be back in about twenty minutes." They went off.

"Oh, crap!" Joe exclaimed as the rain started again.

Chapter Twenty Nine

Rusty was sitting on the floor, leaning against the wall. Tully and Red sat on either side. Mario was just outside the semicircle next to Arron. Red finished briefing Rusty on Colonel Nelson's update and the Post's current situation.

"That puts us in a horrible place, doesn't it?" Rusty was noticeably better, thinking clearer.

"Our options are limited, Rusty. Stay and delay them for a few hours and die, or abandon the Post and hope to slow them further down the road, but we live, as they say, to fight another day. Not much choice, I'd say."

"I agree, Tully." Said Red. "If we wait too long, we'll never get out. That goes against our primary mission now. Delay and disrupt, not die for nothing."

"So, we agree." Said Tully. "Good! Now maybe you'll be open to another option that fits our primary mission."

Rusty looked over at Red. Red just shrugged, spread his hands. "I have no idea where he's going with this."

"A similar experience popped into my head while I was thinking about the living or dying options. It happened when the Nazis surprised us with their counteroffensive at the Bulge."

Mario jumped up excitedly, said, "You want to pull off a Bastogne?"

"Sit down, Mario, let me tell the story. Now you know Mario was with me." He said to the group.

"We were on the south flank of Bastogne. The Jerry's were setting up to turn our position. It was night, it was cold as hell, and we could hear'em moving heavy guns in the woods, one hundred yards at our front. My lieutenant wanted volunteers to sneak in and disrupt their effort. We are, after all, Rangers, he said. My squad was picked from the other volunteers. The Nazis weren't expecting any aggressive action from us as they were busy putting the final touch on overrunning our position. So, my squad crawled through the snow in zero degree weather, snuck in between their lines, killed the crews, and knocked out four of their 88s. That screwed up their plans for the morning attack, which then never happened."

Mario added, "That's when our squad nicknamed the crazy plan as "Pulling a Bastogne." We did it one other time, too, also a success."

"How many men did you lose on that mission, Tully?"

"I lost three of my men, Rusty."

"But he saved our company!" Mario added.

"So, what's your idea, Tully?"

"Ben told me about a hunting story that Arron told him. He'd found a hidden way through the mountain on our west flank. Arron, tell us about this opening and where it leads."

"Wow! Sure. This was a long time ago, just after we got here, I think. Anyway, I was chasing a small deer through the bushes about a mile southwest of here. It darted into this cave at the base of the cliff. I followed, thinking it was trapped but discovered the cave opened into this cut that went right through the mountain to the northside where it opened to the field just beyond the tree line on our western front. You really have to stumble upon it because it's well hidden by brush and the cliff rock. It's narrow but passable."

"Let's check it on our map." Tully pulled out the Arial map book. He pointed at the field behind the trees. "So it opens to this area, right?"

Arron nodded. "Yeah. The opening is about here," pointing to near the middle of the field.

Tully stood, started to pace while talking.

"This is where they're setting up these killing mortars. The beginning edge of the field is about one thousand yards from our trench. This area is around two thousand yards deep. I guess they'll set up these mortars about two thousand yards from our position. With the mortar range of three thousand yards, that'll give them a great angle for the shells to hit every part of the Post. So they'll set these up in the middle of the field, and why not, they have no danger of being attacked. Everything is in front of them. They'll have no rear-guards, and they certainly won't be expecting any close-in attack."

The men were silent. Tully sat back down on the floor.

"So, we'll go in silent and kill them all."

An intake of air.

Silence.

"You'll need most of the men," Rusty said.

"Thanks for your vote of confidence."

"I must be still shellshocked." He said with a slight smile.

"Count me in! I wouldn't miss another Bastogne for anything." Mario was excited.

"You can't go. You're still not right. That head wound has jumbled your brain, Mario."

"Oh shit, Tully. You of all people should know my brain has always been scrambled. Now it just hurts a little. I'm going! I'm not going to wear a helmet anyway. I won't aggravate anything, and besides, Snake and I make the best assassin team that's ever been seen."

"You make a good point. What did Marty say?"

"Says I'm crazy but thinks I'm okay."

"Sure you're okay to do this?"

"I swear to God and hope to live!"

"Alright. You go. Rusty, any other thoughts about the plan?"

"It's risky, but I like it. Red, how do you feel about it?"

"Same. Not much choice. I think it can work, buy us more time."

Rusty said, "What about keeping this cut secret? I wouldn't want them to find out about it. It could lead to the discovery of our escape tunnel."

"I thought about that. I would plan to work our way to the rear and north of their position and then attack, disguising our retreat if we're spotted, so that they think we came in from the northwest. Of course, we'll utilize stealth throughout the mission. Hopefully, we'll get out clean."

Red leaned in, "I suggest we have a backup plan in case things go amuck, and we have to make a hasty retreat under fire."

"I think your right! Any ideas?"

"Larry is magnificent at rigging explosives, he, Ray, and I rigged all the explosives here and on the road. And God knows we have plenty of C-4. So I think we should rig this cut so that when we come back through, we blow it closed. We'll never use it again, so it doesn't matter to us if it's sealed, plus we eliminate the threat of them coming through."

"Brilliant!" Rusty exclaimed.

"I like it too, and it allows us more tactical flexibility. Excellent, Red." Tully added.

"That's why I keep hanging around you whackos. Keep you all from hurting yourselves."

"And it's much appreciated from some of us who would be the hurtees!" Arron's final comment brought some smiles.

Rusty, still not thinking entirely clear, didn't want to make the next decision. "Who should stay, Red?"

"I think Tully and I need to make our plans for this Bastogne first. Tully you ready?"

After talking it out for thirty minutes, going back and forth on options and possible countermeasures, they finally set on a plan. Then figured out teams. The rest of the men sat in silence, listening to their logic.

Red turned to Rusty. "You heard about our plan. Think we need to take the chance to leave the Post almost naked. Don't see them attacking until the mortars are in action."

Rusty nodded, "It's a good plan."

"Good. Rusty, Marty, Tony on the mortar, and Dean will stay."

"Let's assemble the men and fill them in on the plan. We need to move out." Said Tully, now the leader of this visitation to Bastonge, as he and Red walked out of the room.

Chapter Thirty

They kept only the essentials; no helmets or anything loose that might make noise, just killing things. Snake guided them down and through the long escape tunnel to the exit hatch, then up to the outside. No sign of the enemy as they exited, proceeded along the cliff edge through dense ground cover shrub and trees. Arron took the lead with Snake behind him, Tully was next in line. Larry brought up the rear of the column.

Ten minutes into the trek, Larry, who hadn't brought his rifle or Tommy, because he was carrying so much C-4, suddenly stopped. In one motion, he drew his .45 shoulder pistol and spun around.

"Don't shoot!" Marty said loud enough for Larry to hear him.

"Jesus Christ! What the hell are you doing here? Does Rusty know you're here?"

"Sorry, Larry. Thought I'd be needed here more than at the Post, just in case."

"Ah, shit! Okay, you're here now. Get moving. Catch up with Tracker upfront."

At the front of the column.

"Are we close?"

Arron turned to Snake, "I have no freaking idea. I can't see shit, and I haven't been near here in a while. Like I said, we're going to have to stumble on it."

"Well, hurry up and stumble."

The rain came steady and hard. The column kept moving.

"What the hell!" Arron said as he fell and landed in a small stream.

"Found it!"

"You okay?"

"Wasn't expecting the damn Rio Grande! Yeah, I'm good."

Arron followed the stream to the crevice; water was pouring out from the cliff runoff. He hesitated.

"No time like now, Arron, you won't drown." The water was six inches deep and running fast as Tully pushed him into the cut. Arron continued the lead through the opening.

There was banging and scrapping as the men wiggled through the narrow cut. The machine guns were the hardest to keep from hitting the walls.

Larry went in last. His backpack was full of C-4, and he soon had to remove it from his back to maneuver farther in. After about thirty feet, he stopped, got his flashlight out, and started scanning the walls for a place to plant the charges. The runoff coming down the walls was intense. Standing knee-deep in water, he finally spotted a ridgeline about five feet off the floor. It had small mini caves in the walls that ran about six feet long. Perfect, he thought. He went to work setting the charges.

The men filed out of the crevice into dense foliage, making their way west as planned. With Arron still in the lead, Tully moved in front of Snake, as they moved mostly by hand, feeling their way, making some noise banging into trees and rubbing through the bushes. The rain hid the sounds. As the brush thinned out, they could see a line of dim lights in the center of the open field stretching north.

Tully tapped Arron. "Hold up." Turning to Snake behind him, he palmed him on his head to indicate that they'll take cover here.

Snake did the same to Mario behind him and repeated it, and the men went to ground.

Tully positioned his walkie-talkie and clicked the send button three times. He silently prayed. Please, God, whoever you are, let me be right about this. I know, I know, I've begged you for help so many times. Just one more time, okay?

Tully heard two clicks in return.

At The Post

Rusty was in the mortar pit with Tony waiting for the signal when it came.

"Okay, Tony!" Tony lifted the illumination shell and dropped it into the tube. He had positioned the mortar so the flare would hit over the same general area they had previously used; the open field between the woods.

Rusty said a prayer. Please, Jesus, keep my men safe. And oh, please don't let the enemy attack the fort right now. Thanks.

The Field

The area Tully faced became cast in shadow light as the flare ignited, clearly highlighting the trucks and men setting up the mortars. The enemy just looked up at the brightness. It was about a thousand yards away, so what, they must have thought.

Tully smiled. Thank you, God! It was, as he had envisioned. Twelve track vehicles in a row, parked behind mortars spaced every thirty feet or so, in the middle of the field. No sandbag emplacements; they were totally unprotected. Men were stacking shells from the tracked trucks in front of the mortars. The field must have been a pasture at one time, he thought, gone wild now, with large bush and undergrowth. Perfect for their approach cover.

The men started fanning out to their assigned areas. Tully had gone over his plan with them before they left, hoping that he had figured the layout of the enemy correctly. After seeing the field lit up for thirty seconds, he guessed right, and they knew how to proceed.

185

Red and Arron were moving out from the bushes when they heard a whisper behind them, "Hey, wait up!"

Red stopped and turned; Marty bumped into him. "What the…"

"Marty!"

"Please, Red. I've already been chewed out by two people. I'm here, let's make the best of it, okay?"

"Goddamn right! Come on! Take our right flank."

Red and Arron set up their machine gun between two small boulders and a patch of shrub bushes. It was thirty yards to the east and south of the mortars, close to the cliff rocks. It gave them an excellent field of fire against any troops which might respond from the tree line. They could also shift the gun and rake the mortar line if need be.

"Think this a good spot, Red? I can't see the tree line."

"Good! Then the enemy can't see you either. Get the belts set." Arron dropped the five loose ammo belts from his neck and loaded one for Red. Red locked and loaded it. Click. Click.

Red turned to Marty. "I'm very pissed at you, and I'll talk to you later, but for now, keep your rifle up and head down. And if I hear a peep out of you, Marty, I'll knock your teeth out."

Mario and Snake belly crawled the thirty yards to the nearest mortar. The mortar positions were easy to find. The NK troops were working by kerosene lanterns, not a lot of light but perfect for what they were doing.

"We need to wait for" Mario checked his watch dial, "four more minutes. I'll get any guards around the truck, definitely one in front, maybe another in back. You've got three men around the mortar, will you be okay?"

"You've never seen me throw a knife. I carry three. I'll be fine."

Well off to the rear of the trucks, Manny and Ben were running north across the field.

"Come on, Ben! We've got like three minutes."

"Shut up! We'll be there in a minute."

Ben could see the faint glow of the lamp at the last mortar position, the one most north. Tully worried that reinforcements might come from the open area to the north. Their job was to act as blockers should any such force show up. Tully also thought that area was probably a staging area for the attack on the Post, and troops might be deployed nearby.

Manny had no illusions about their job here. He knew that when the shit hit the fan, it was going to get hot as hell here. Please, Lord, let me live, he prayed.

Manny stopped thirty yards from the mortar on its north side, found a low hedge for cover, and set his machine gun. Ben followed, removed his ammo belts, and set the tripod.

"I told you! We got a minute to spare! Take a nap."

When they approached their assigned mortar, Joe and Frenchy realized they had a problem. The third mortar from the north. Their task was to take out the first three mortars. They hadn't seen the tent to the rear of this position and almost crawled into it. When they were within five feet of the tent, someone lit a kerosene lamp. They heard voices and saw shadows from within. Two, three, they couldn't tell.

Joe whispered to Frenchy, "Must be the company HQ." Frenchy whispered back, "You think he's a lieutenant or a captain?"

"Is he what? I have no freaking idea! Who cares?" Joe was puzzled.

"We've got to take these guys out before we start. This could blow the whole deal if we screw up. I'm trying to think of a way to get them out of the tent without alerting everybody, any ideas?"

"Make a sound; they'll come out to check it."

"You're an idiot, Joe. It's raining like hell. They'll never hear shit. Besides, they'll probably come out guns ready. I've got an idea. You take the north side of the tent flap; I'll go to the south side. Knives only. Be ready to rush in."

"You sure you know what you're doing?"

"Hell, no!"

They moved into position. Frenchy stood up to the south of the tent flap, and in a loud voice, he called out in Korean. "Sir! We have a problem out here!" A response came back in Korean. Frenchy shot back in Korean, excited. "No, sir! You must see this!"

They heard grumbling and movement. The flap opened casting light on Frenchy. The officer was startled. Frenchy grabbed him, shoved his knife into his throat. Joe darted quickly into the entrance and slashed the throat of the other soldier who was standing, empty-handed, with a total look of surprise on his face. He turned, fell to the floor, gurgling, kicking his legs, and then he stopped. No one else to kill here thought Joe. Good.

Frenchy slipped in behind Joe, "We got more killing to do, my friend. We're behind schedule. Move."

Beyond the two flanking machine gun cover teams, Tully had divided his people into two-man hit teams. That left eight men in four teams to destroy the mortars. Not knowing how many mortars there would be, he assigned the teams from north to south and told them to divide the total number of mortars they found by four. Starting from the north, they'd then count down. Any odd number would be left to the furthest south. They would attack from south to north.

Tracker and Foxy were set to the rear of the sixth mortar. Moose and Tully were behind the ninth mortar.

Tully glanced at his watch; thirty seconds. Damn! He thought. Five years later and here I am lying in a mud field, soaking wet and getting set to kill. Or will it be me that gets killed this time? You have been a lucky son of a bitch, haven't you? But you got a lot of men killed, didn't you? Stop it! Remember all the men you saved, you idiot! Focus, focus. Focus!

The teams crept into attack position. Silently, they started doing their thing. Throat slashed, neck broken, stabbed through the heart; the mortar crews began dying in numbers.

Mario was behind the third truck. The guard was just standing there, trance-like. Maybe he's sleeping standing up, he thought, I've done that. He cut his throat and laid him down. When he turned toward the rear of the truck, a soldier was standing there, he hesitated for a second then turned to run. Mario reacted, throwing his knife into the soldier's back. He yelled out a scream and fell. It was enough to warn the two remaining men by the mortar. Snake was lightning fast, seeing the reaction to the cry, he let his knives go in three seconds flat. End of that little trauma, Mario thought. He went small, looked to the next mortar to see if anyone was alerted. Of course not, he thought, they're all dead.

Frenchy was on their third mortar as well as he approached the guard leaning against the truck, smoking a cigarette. Joe was taking care of the crew to the front. The driver turned suddenly to him. "Who are you?" He said in Korean. He couldn't see Frenchy. Must have just sensed his presence. In Korean, Frenchy replied, "I send you to your departed family." He thrust his blade into the soldier's throat.

Foxy was perplexed as he and Tracker started on their third mortar. He had taken care of the last two truck guards, but when he'd gotten to the mortar crew, they were all dead. He wondered. How could Tracker take care of these guys so quickly? Now he decided to move as fast as possible, so he might help if need be, but really to see how Tracker was doing this. He got his man fast and ran to the mortar.

One man was already slumped on the ground. Foxy saw Tracker whirl over his body and sink his tomahawk into the second soldier's head. Then, in one motion, spun in the air and buried his hatchet in the last man's neck. Jesus! He thought I've never seen anything like that! I didn't even know he had a tomahawk! Man! These injins are really dangerous!

On the not too distant rise to the north, they saw vehicle lights approaching. Ben saw the lights first. He nudged Manny, "We got company coming." As they came over this hill, Ben was counting, three, four, five, "Oh Christ, we got trouble."

"Trucks, maybe? Troops? Supplies, I hope." Manny reached for optimism. "No matter. We can't let them get too close. We can't run

189

until they start blowing up the mortar tubes. Guys! Hurry up; God damn it!"

"Calm down, Manny! We gotta open up in about thirty seconds."

Tully and Moose were on their last mortar. Tully had taken out two of the crew when a third member he hadn't seen, appeared. He fumbled drawing his gun. Moose shot him in the head with his pistol. He saw the last crew member turn to run, so he killed him as well. The shots were loud and distinctive in the rainy night.

"Thanks, Moose. Think you woke up the neighborhood."

"Sorry! Can't throw a knife."

Tully saw the lights coming their way. Then tat, tat, tat …..
"Shit, that's Manny shooting!"

"Set the cab, Moose! We gotta go." Tully dropped a grenade in the mortar tube as Moose dropped one in the truck cab. They were moving fast now, going down the line. Two dull booms went off as they started on the next mortar and truck. They heard other explosions as the teams executed the plan.

A few minutes earlier

"Now, Manny!" He fired at the lead truck, raking the cab first. One light went out, then the truck veered sharply to the left and stopped. He moved his fire to the covered canvas rear, then on to the next truck. The third truck immediately turned to the left, trying to pull behind the second truck for cover. The other trucks did the same. Manny kept firing. Ben heard an explosion.

"They started! Let's go, Manny!"

"One minute, I'm not finished." He kept firing, more explosions. Troops were jumping out of the trucks and started to return fire.

Joe and Frenchy were running from their last burning truck into the field behind it when an illumination flare went off four hundred feet above their head; they dove to the ground.

"Crap! Look at those guys! Do they have a death wish?" Joe was looking at Manny and Ben two hundred feet away, still firing. It was almost like daylight now. He yelled at them. "Hey! Assholes! Let's go, come on!"

Manny was on his last belt when the flare went off, then ran out of ammo. He heard yelling behind him. "What did he call us?"

Ben grabbed Manny's arm and pulled him away from the gun. "Let's crawl to the guys." They started, with gunfire getting intense as the enemy soldiers fanned out from the trucks. The flare died, Ben and Manny got up and started running. As they got to Joe and Frenchy, another flare lit up.

Manny cried out and fell face-first into the mud. Everyone went prone.

"You hit?"

Ben turned Mammy's face a little, "Oh, man, that hurts."

Both Joe and Frenchy, still flat, opened fire on the advancing troops.

"Joe, when the flare goes out, help me with Manny. We need to get out of here."

Joe crawled closer to Manny, and when the flare extinguished, he slung his Tommy gun on his shoulder and grabbed Manny under his arm as Ben grabbed the other. They half-ran, dragging Manny, his feet bouncing through the muddy ditches. Frenchy ran behind them, covering the enemy troops. The burning trucks cast a dim light into the field, enough to guide the fleeing men, but it also put them in silhouette.

Tully went flat, ten feet behind his last truck. He knew his men were in trouble. Well, he thought, so much for an easy exit. I owe you one, Red.

When the second flare ignited, soldiers started rushing out from the tree line. The fiery trucks were like honey to ants. They started firing in that direction. Red was waiting for them to get closer. He thought fifty yards would do it. They were close now. Maybe sixty men. Won't be that tough, he thought, I've got good cover.

"Shouldn't we open fire Red, they're getting kind of close?" Arron was nervous.

"Good idea." He started on the closest soldiers, moving the firing lane north. Most of the troops were running and firing, focused on their mortars and trucks burning

. They didn't see or hear Red's machine gun chewing up their comrades on their flank. When the flare extinguished, he stopped.

Red had seen Marty sprint off out of the corner of his eye when he was firing. What did he see?

Arron hit his arm and pointed towards the mortar line. Enemy troops coming down from the north had reached the farthest mortar placement, started moving toward their exit point. Weren't many left from the woods, so Red shifted the weapon over to these troops and opened fire.

Ben's group reached Tully in the field and put Manny down, then lay flat themselves.

"Who's hurt?"

"It's Manny; got hit bad!" Ben called out.

"Moose, get Manny on your back. Ben, Joe, give him a hand."

Moose knelt, offered his back. Joe and Ben lifted Manny onto the big man, arms hung loosely around Moose's neck. He grabbed Manny's arms and stood up and took off toward the cliff. The enemy was getting close. Still prone, Tully and Frenchy opened fire, then heard Red's machine gun firing and saw some of the attacking soldiers fall. Then the rest of the pursuing troops went prone.

The flare died. "I don't think we need an invitation. Let's skedaddle!" Tully and Frenchy rose together and ran.

Marty had seen some of the action in the field, knew someone was hurt, so he made a beeline back to the exit point. He was prepared when Moose laid Manny down. Moose had a shocked look, seeing Marty, like everyone else who came in, but also one of relief.

Manny was breathing hard and covered in blood, front, and back. Marty was working fast, plugging the holes, wrapping them tightly.

Red and Arron got to the cut at the same time as Tully and the rest of the team. Marty was just finishing getting Manny ready to travel.

"Marty!" Tully was rattled. The enemy was close behind him. They needed to move. His man was hurt, and now Marty shows up from nowhere. "Jesus H! I'm not sure if I should kiss you or shoot you right now, I'll decide later, how's Manny?"

"Not good, Tully. The bullet went through and through, which is good. Going through the lungs, I think. He's wheezing and coughing up blood. I got the outside holes patched, and blood loss stopped. But I don't know what's happening inside? I'll have another go at him at the Post, but he needs a surgeon. Tully, I mean, he will die if he doesn't get to one real soon."

"Shit! That's gonna be a challenge, Marty. A real goddamn problem!"

"Red! Help Moose get Manny through the cut. It's gonna be tight. Try not to bang him up too bad, but speed is more important than some bruises. Okay, let's go! Frenchy, bring up the rear, I'll be in front of you."

It stopped raining.

Larry waited anxiously outside the cut. The gunfire and explosions had echoed over the cliff, telling him all had not gone as planned. How many made it? Did anyone make it? Only three guys plus me, left if they hadn't.

He heard sounds in the cut, scraping, and banging. He moved away from the opening, knelt by the explosive plunger he had wired up and waited.

"Larry, it's us! Don't blow us up, please!" It was Snake yelling at him.

He yelled back, "Sounds like you had enough excitement for the night."

Snake came out, "Yeah, too much fun, alright. Didn't want to leave."

Larry wanted to know about casualties but couldn't ask the question. He didn't want to spoil their or his relief. The men filed out; he saw a man being carried. Who was it? Too dark to see. Only one. Shit, I didn't count the others. Did we leave anyone?

Moving slowly, the men followed Snake back to the tunnel.

"We're it, Larry. The next guy through won't be ours, so get set!" Frenchy and Tully moved away.

They got fifteen feet along on the side of the cliff when Larry pushed the plunger. Good thing too, because the narrow cut acted like a gun barrel, directing the explosive force out into the bushes and trees for ten feet. Rock rained down in an arch around the entrance.

Tully looked back at Larry. "You think that's enough to block the passage?"

Larry did a double-take. "You're kidding, right?"

When they arrived, only a few men remained waiting at the shaft entrance. Manny had been one of the first to go back inside.

"Snake and I will close up and make it look beautiful." Said Foxy. "We'll go around and up the road. Please tell whoever's watching not to shoot us."

"Yeah, got ya covered. You be careful, Foxy, you never know what you may find around here now." Said Tracker, as he climbed down the ladder to the tunnel and secured the inside hatch.

Foxy slid the hatchway into position and made it look natural again. Snake led, as they worked their way around through the brush to the road. The night was still black, with low clouds threatening more rain. Like a sailor following the stars, Snake had an internal guidance system and knew every inch of this route. They were silent as they moved up the road towards the gate entrance. Of course, the gate was no more. Shell craters pot marked the surrounding area.

At a hundred feet from the Post entrance, he stopped and listened. Hearing no danger, he started to crawl than stopped suddenly. Directly in front, he heard a scraping sound, like cloth on gravel. Foxy heard it at the same time and froze. More noise. It sounded like one person.

Snake moved right to circle the sound. Must be one man, he thought. Foxy was calm and slowly drew his .45 from its chest holster. What now, he thought. Difficult to imagine the enemy could get here? In the dark. And how could they find the road entrance? I guess we'll find out.

"Oooooh.......Foxy, Foxy!" the man cried out. Then in Korean, "I am a friend, from Captain Hui, need Foxy." Snake jumped him and had a blade at his throat.

"Easy Snake, I think he's okay."

Then in Korean, "Are you alone?"

"Yes, we need Foxy."

"I'm Foxy. What do you need?"

"Captain Hui says we need your rockets to kill tanks; we're out."

"What's your name?"

"Lieutenant Lee Kun, I came in with new company yesterday."

"Alright, you're coming with us, don't make any strange moves, or I'll kill you. Understand?"

"I am a friend, I make no bad move!"

"He's good I think. Check him for weapons. You lead us in. I'll cover him."

Chapter Thirty One

"He's alive and breathing, well, wheezing. He's in a terrible way."

Rusty and Tully were kneeling next to Manny, Marty was sitting on the floor, slumped against the wall.

"Let's talk about you for a minute." Rusty was upset, not just about Manny but Marty as well.

Marty was also agitated. "Hear me out first, will ya! I know what I did was wrong. I'll never do it again without putting up my side of the argument first, and I swear, I'll abide by whatever you decide. I should have done that, but I didn't. I firmly believed that the enemy wouldn't attack the Post until they had those mortars ready. I felt I wasn't needed here, but that I was truly needed with Tully."

"I say we shoot him." Red barked as he walked in. He had heard most of Marty's little speech.

Tully shifted to a sitting position, "You saved Manny, no question about it. But you put the Post in jeopardy, and you broke trust."

"I know you, Marty. This is not like you. Why didn't you say something?" Rusty pressed, still troubled.

Marty seemed to weigh something in his mind. He looked at Rusty. "Something like this happened to me on Okinawa. I had this feeling, just like here. I wanted to go on this mission but was ordered to stay. I wanted to go anyway, but I didn't act. Lack of courage, no balls! I don't know why! But my strong feelings bore out, and many men died. I have deeply regretted my inaction. I thought, and still think, that I could have saved a few of those men. I didn't want that to happen again."

"Next time you get that feeling, you come talk to me. I promise I'll listen. Thanks for saving Manny tonight." Rusty knew about those feelings, understood Marty's need to go.

"How much time does he have without a hospital?"

"I don't know, Tully. All I know for sure is that it's not long, six, twelve hours at most. Even then, a hospital may not be able to save him, but at least he'd have a chance."

"Okay, let's talk about some options. Red, how long can we hold out?"

"Another day, max. I think the Ridge will go first. Then they'll be able to surround us. Come at us with everything, more tanks, and enough men to overwhelm us."

"Tully, let's see what Foxy can get from that ROK lieutenant before we make any major decisions. Maybe we have more options than we realize. I also want to call Colonel Nelson; he might be able to help, and we certainly need to know the latest situation."

"It's good to have you back, Rusty. We really need your clear thinking again."

Foxy came in, sat on the floor. "How's he doing?"

"He's alive, but we need to get him help. What did the lieutenant say?"

"They've taken a beating but are still hanging on. Says the *Cannon Jeep* and bazookas saved their asses. The bazookas knocked out six tanks, but they lost the jeep and one bazooka, and finally ran out of rockets. The next tank assault will run them over. Captain Hui thinks they'll resume their attack this morning, at first light." He looked at his wristwatch. "In about an hour."

"How many men does he have at the Ridge?" Did he say anything about getting more reinforcements?"

"They have less than a hundred and fifty men that still can fire a weapon and lots of wounded. They were told not to expect any

197

more help. Their fallback position is the Streambed, that's being heavily fortified. The rest of their regiment is taking up positions there. That's why Captain Hui is trying to hold out as long as possible, give them time to prepare."

"Do they have artillery? Any anti-tank weapons?"

"That artillery unit that helped us at the bridge was almost wiped out. What was left moved about five miles farther south and been folded into another battery."

"That explains why we couldn't contact them," Rusty noted.

"Right. To finish up, the Ridge has run out of everything that can stop the tanks except for some homemade wired devices they laid in the road and laid in the approaches to their front in the open fields. He says that Captain Hui learned this from our guy when he was preparing those IUDs in our road out front. I think he means, Ray. So that and the mines we gave them is all they have to try to stop the tanks."

Rusty smiled, "I didn't even know that Captain Hui knew about those devices until you told him, Foxy."

"Yeah, me neither. Ray must have thought Captain Hui was pretty unique, trusted him."

"Well, he got that right!" Rusty looked up to heaven. "Thanks, Ray. You're gonna save some guys soon." He turned to Tully.

"How many rockets did you bring with you, Tully?"

"I have no idea. We had stacks of all kinds of ammo boxes. I know we brought two bazookas."

"I know!" Said Marty. "I'm the local mule," he said with a smile. "I unloaded most of the crates and took inventory."

"So how many?"

"Four crates, six per box that we haven't used. We still have rockets from our initial shipment, twelve or so left."

Rusty stood. "I think we need to get those rockets to the Ridge, and we need to get Manny out of here. Any ideas?"

"Yeah, let's call Colonel Nelson." Said Red.

Japan - Eighth Army Headquarters – Colonel Nelson's Conference Room - 0500 – (5:00 am)

Colonel Nelson's modest conference area was now a war room. Four men handled phones at a makeshift table near the room entrance. Additional chairs had been brought in and placed around the large oval conference table. A couch appeared in a corner, and cots set along the walls. The conference table held stacked folders in several spots and strewn about its expanse were old coffee cups and plates of stale leftovers.

His staff was scattered around the room. Some sat at desks made of stacked crates. Others sat near a phone bank on a corner of the conference table. A similar phone bank faced Colonel Nelson at the other end. The table phones all had a light on the side, and tabs taped to the bottom front. There were three red phones, "G2 Seoul" on one, "Kimipo" and "Post." These were direct connections from the Korean cable. The white phone connected to General MacArthur's office labeled "G M." The other three were black, labeled one to three.

Maps, large and small hung from every wall. It had become the unofficial Eighth Army "War Room." Colonel Nelson's new home since late yesterday.

Nelson had only been on the phone with Rusty for a few minutes. He heard the urgency in his voice, about Manny and the Ridge. Rusty wanted guidance about the Post. Stay or leave?

"Rusty, listen. I won't order you to do either. It's your call. Your efforts so far have been heroic, way beyond what I envisioned. What happens from here on depends on events beyond your or my control. I will say that you're buying precious time. The MLA (main line of attack) has been stopped cold at your defense. The ROK Army is frantically building a second and third defensive line south of you, and it looks like the US of A may join the fight very soon. But you know how the wheels of Government turn. Nobody is counting on anything happening right away." He gathered his thoughts before continuing.

199

"Do what you can to help the South Korean's on the Ridge; because you're right. What happens there determines what you do at the Post." Pausing again to build out an idea he just was forming.

"I have a thought about getting Manny help. See if you can get him to the ROK unit at the Stream, I'll see what I can do to get him out from there."

"Colonel! You must still be talking to that Leprechaun? You work this miracle for Manny, and you'll start reaching Godlike status."

"Bullshit aside, Rusty. You're now going into deep shit. Whatever happens, I'll try to help."

"Appreciate it, sir. But remember something, you may have recruited each of us, and you're a good salesman, but each of us signed up. We're here because we wanted to be. Now please, don't forget to ask for tanks."

"Tanks! Yeah! I'm working on that. Call me in an hour if you can; good luck."

"Hey, wait! Have you heard from Captain Jordan? Is he at the Peak?"

"I haven't heard from him. I haven't heard from any of the Posts."

"Shit!" Said Rusty and ended the call.

Next, Nelson picked up the red phone labeled "Kimipo."

"Corporal Andrews here."

"This is Colonel Nelson. That was a pretty quick pick up corporal. Why are you at the phone at five in the morning?"

"Sir! We got the word to prepare for an evacuation at midnight. All hell is breaking loose here, sir."

"What evacuation?"

"Colonel, we were notified by General MacArthur's staff that all US nationals had been ordered out of Korea. Our embassy sent over the list of eligible personnel."

"Oh, Christ! How many names are on the list, corporal?"

"I didn't count them, sir, but Billy said there were over two thousand."

"Where is Billy? How many men are with you at the base?"

"Billy's on the flight line. There are only me, Billy and the loader, and two pilots who are sleeping."

"Great!" He was dismayed and angry.

"I called to find Captain Bill Stands. He's a pilot. I need to talk to him ASAP."

"Stands? Right? Oh! I know! You mean *Wild Bill*!" Sure, he's here. Do you want me to wake him?"

"Yes! Tell him it's urgent. I'll wait."

As he waited, he thought of this new development. Why am I surprised? The people who really know what's going on are never consulted. Confusion and chaos; the world is mad! That's why I was put here. Right!

"Is this my favorite Colonel? Waking me at five AM? This must be a real doozie!"

"Yes, it is. I don't have time for niceties, so I'll get right to it. I need you to fly into a battle taking place this morning about four miles south of the border and rescue a severely wounded American Ranger. It's not an order. It's a request."

"You make it sound so fun! Too exciting to turn down, Colonel! My blood is flowing. Tell me, are there enemy fighters involved? And OH!, is this Ranger on a stretcher?"

"No sightings of enemy aircraft so far, but no guarantees either. Yeah, he's on a stretcher, chest wound. You got anything that could work?"

"Can't land in a field with the rain we've had, so a road will have to do.

"That also limits my selection to the one plane I have. It'll take a stretcher after I remove a back seat."

"So you'll do it?"

"For you! Anything! I wouldn't miss this for the world!"

Colonel Nelson gave him the details and established a time-line then ended the call.

Then he grabbed the white phone.

Korea - The Post – 0515 – (5:15 am)

Tully, Red, and Foxy were sitting in the corner with Rusty. Marty was with Manny on the back wall. You could hear his labored wheezing.

Rusty began. "A few hours ago, I thought we found a temporary solution to our dilemma but didn't anticipate just how temporary, temporary was." Pausing a second. "It's now time we plan our exit. The next push will overwhelm us. According to Colonel Nelson, we're holding up the whole main line of attack. So we've done our job. Putting myself on the other side, I'd be really pissed. Pissed so bad that I now wouldn't care how many men I lost, because the prize of capturing Seoul and then all of South Korea is worth almost any price. So I think the next attack is it, and it'll probably come at dawn."

Red jumped in. "Rusty, we can hold out for another day and escape tomorrow night."

"You may be right, but I don't think the Ridge can survive another day, probably not even another couple of hours. I'm not sure we can either. From what that lieutenant said, they have a hundred and fifty men left behind an unarmored earthen berm. Their minefield is mostly gone. The stupidity and arrogance of the enemy are gone; they won't repeat the same mistakes. But most of all, the moral of the South Korean troops are shot. They've been hit hard, and there's too much stacked against them. Their Captain Hui must know this."

Red nodded, "You're right about the minefields, and I mean ours as well. The constant shelling has taken out a good deal of our mines. That will leave us very vulnerable when they next attack in force. But that's a moot point if the Ridge is overrun."

"Right, so here's what's going to happen."

"Aren't we going to discuss this, Rusty?"

"No, Tully, we're not. Like you said earlier, I'm thinking clearly now."

"We're leaving the Post! I see no alternative. Their next attack is for keeps, and nothing is going to stop them. So listen up, we're not going to make this easy for them."

"I want Ben and Frenchy to take Manny to the ROK position south of the Stream. Tully, will you see to that? With any luck, Colonel Nelson will figure a way to get him out of there."

"I'm on it." He turned to leave.

"Hey, Tully, I meant to tell you. That was one hell of a **Bastogne!**"

Tully smiled and left.

"Red, I want you, Foxy and two more men that you choose, to take our four crates of bazooka rockets to Captain Hui. Take three bazookas, give them two. You keep one."

Red nodded, "I'll take Joe and Tracker, but I'd like two more men. Tony and Mario would be good. What do you think?"

"You're right; that's a lot to carry."

"You might have to stay at the Ridge if the attack opens at dawn. If it doesn't happen, you get out of there. If you can't, hold off as long as you think safe, then fall back down the road to just south of our compound, where the woods are on either side of the road. In either case, that'll be your rear guard position. Convince Captain Hui to do the same and move his troops out down the road. We'll cover your retreat from here for as long as we can, then follow through the tunnel. We'll meet south at the road where we had planned. Make sure you explain to Captain Hui that we are withdrawing. It will leave his position untenable. He must withdraw with you."

"Take one of the walkie-talkies. Frenchy will take one, and Larry will keep ours."

"Foxy, make sure Captain Hui notifies the unit at the Stream about our guys bringing Manny. Press him on our retreat plans."

Tully came back with Ben and Frenchy. Rusty filled them in on his plan. Then they got Manny set up to leave. Marty had rigged an IV pole to the stretcher. He wanted to go with them, but Rusty said no. With faint illumination from nearby explosions, Ben and

Frenchy bore Manny on a stretcher, out of the building, down to the road, and began their mile-long walk to the Stream.

Ben was having a hard time dealing with Manny being shot. It reinforced his feelings of being a jinx to his friends. On their first break, Ben told Frenchy about his hex on his friends, that they die, and how Frenchy should get away from him as soon as he could.

"I've known you for some time, my fellow Ranger. And I've never known you to be bad luck. So, you think, maybe it's the water?"

Ben was silent for a moment, "Frenchy! Maybe you're right."

Artillery fire had never actually stopped. Throughout the night, a constant drumbeat of explosions surrounded them; it just wasn't intense or concentrated. That was Rusty's primary concern now, a renewed heavy concentration just when they were out in the open.

"I've got everybody loaded to the gills. We're ready to shove off." Red looked around. "I hate to say goodbye to this place. It has good memories for me. Has Larry prepared the charges yet?"

"Yeah, Red, he has. I feel the same way. You did a great job here. Good luck over there, see you down the road."

Red led his men down to the road where Foxy and the ROK lieutenant then took the lead.

Tully turned to Rusty, "What charges?"

"You didn't think we would just abandon this compound intact, did you? Without maybe leaving a little surprise for the new occupants? From the very beginning, we planned the end. Throughout the building and under the trenches is enough C-4 to blow up a good size mountain. The detonator is at the tunnel exit. We'll have a clear view of the building from there. Should be quite a show."

"Tully, get our guys ready to go, then get back to watch. I want the rest of any bazooka rockets loaded into backpacks; we are taking our last bazooka and a machine gun you brought. I'll be with Larry."

"Do we need to take extra ammo?"

"I thought I told you about our depots. We have plenty of everything at each one."

"Sorry, you did, just forgot."

Chapter Thirty Two

When they arrived, Joe told Red he had scouted the Ridge the other day and wanted to establish their positions close to the road, as he thought this was the most logical main line of attack for the enemy tanks. Joe was surprised when Red just turned to him and said, "Joe, I trust your judgment, get our guys set up. Leave me with a bazooka in my position and get the other two weapons to people who know how to use'em, ours, or ROKs."

Four ROK guys appeared as soon as Red left. They were jabbering to Joe and pointing at the bazookas on the ground and nodding enthusiastically.

Joe thought about Foxy's Korean cheat sheet again. He improvised. He signaled to one guy to pick it up, then showed him a rocket.

The soldier nodded a few times, smiled, then picked up the weapon. He said something to the guy next to him, who immediately got the rocket and placed it in the tube and wrapped the arming wire, then taped his buddy on the head as a signal to fire. The second two nodded and did the same. When finished, both teams held the bazookas. There were grins all around.

Joe realized these guys were trained on the weapon, then nodded to them and gave each team rockets. They said something, bowed, and left.

He hadn't noticed two more soldiers to the side that had brought their bazooka and were pointing toward the rockets on the ground. Joe gestured to them to take the ammunition. They did and also bowed and left.

Joe smiled. Now we got four bazookas on this ridge. He had this strange sensation, God-like almost. He felt like Santa Clause, giving out gifts, good killing stuff.

Meanwhile, Red and Foxy were talking to Captain Hui. They told him what was going to happen. He was in denial. Didn't or couldn't think about retreating.

Joe was setting up Red's bazooka while he watched the interchange. A burning tire from a mangled truck in the rear shed a dim shadow over the men. There was a heated exchange. Red stormed off, and a few minutes later, Foxy followed. Captain Hui was upset. Joe decided to go over to him.

"Captain Hui, my name is Joe. We haven't met before, but we have something in common."

"This is a bad time for small talk, Joe. What could we possibly have in common?"

"I went through eight years of grade school, having been taught by Catholic Nuns."

Captain Hui looked at him and smiled. "Must have been fun."

"You don't know the half of it. I have to tell you my story." He did.

Captain Hui burst out laughing, a real belly laugh. Joe had his opening.

"You know what you have to do, Captain. You don't want to waste your men here. We'll all get to that second line at the Stream and fight again—no need to stay here any longer. Just look at your men, they're exhausted. They need a break. What do ya say?"

Hui looked around and was quiet for a minute. "I now consider you my brother, Joe. We've been through hell as kids, and now are going through it again. I hope we live to tell a funny story someday about this time. I'll get my men ready."

Joe walked back to his position and came near Red.

"What the hell did you say to him?" Red was pleased but perplexed, thought he had used a very logical approach, but it had only made the captain angry.

Joe kind of smiled and, in a heavy Brooklyn accent, said, "Just talked some good Brooklynese to him, ya know, 'bout Nuns and stuff."

Red stared at him. "Where the hell did you come from anyway?" Not having any idea what he just said.

Captain Hui called his battalion commander at the Stream and alerted him about Manny coming down and that he was starting the Ridge withdrawal, wounded first. He then issued orders to his men.

The able-bodied still manned foxholes and trenches along the natural ridgeline. These men east and farthest from the road consolidated their line and had been moved in closer, anticipating the withdrawal.

The less wounded started helping more severe casualties to the rear, closer to the road. Some of these men began assisting the wounded leave by helping to lift them onto the backs of the stronger soldiers. Others used tents as stretchers. The line of men leaving grew long.

The Americans stayed in two three-man teams near the road but spaced about twenty feet apart. A ROK bazooka team was placed twenty feet east, then another team at twenty feet. Almost fifty riflemen and machine guns filled the gaps in trenches and shell craters. Another fifty or so soldiers held the ridgeline further east. The one thousand yards wide Ridge defensive line had shrunk down to about three hundred yards.

Foxy called Rusty and confirmed their plans. They were set, and the Ridge was being evacuated.

Incoming mortar rounds started hitting the entire Ridge. It was heavy. The exposed men who were leaving started running, the ones closest to the trenches dove for cover.

Dawn had started. The thick cloud cover cast dull gray light along the battlefield. It started raining.

"They're shelling the Post!"

Foxy nodded, "This isn't going to be easy Red, how long do we stay here?"

"Just long enough to put a damper on this attack. If the Commies take this Ridge too early, we'll all be slaughtered."

"Tanks! Coming off the road in the west corner, about four hundred yards, I'd say." Tracker was nervous. He fiddled with his bazooka. It was loaded; he just needed to set the firing mechanism.

"Do you want the bazooka, Red?"

"No! But now is about the time you should fire, Tracker, so get to it."

The tanks started fanning out across the field. Infantry started coming out of the woods to support the tank attack. A lot of men. They were firing, the tanks opened up, and their mortar rounds started exploding around then. The men on the Ridge began to die.

Tracker fired his bazooka first and hit the lead tank. A second hit on the same tank came from the ROK team to his right. The Ridge machine gunners opened, mowing down scores of advancing men.

Within a few minutes, five tanks were on fire, but they kept coming. Joe saw an enemy unit running unopposed toward the east flank, where the ROK soldiers had been withdrawn into the smaller line. Oh shit! Not good. He figured that if they get to their unoccupied line, they'll roll us up like a can of sardines. He turned to Tony.

"Gotta go, Tony. Mario, come with me!" They ran behind the Ridge on the edge of the manned positions. Shells are bursting around them. Farther down, Joe saw a crater with two dead soldiers around a machine gun.

"Mario! Let's get that weapon." They jumped in. Joe grabbed the machine gun, tripod, and all. "Bring all the ammo you can find."

Climbing out, he spied a trench far down the ridge with a firing position intact. They ran, dodged shell bursts, and finally jumped in. Mario flopped in beside him, ammo belts flying everywhere.

Joe was quick to set the weapon as Mario loaded a belt. The charging enemy had made significant progress since Joe realized

209

their intent, so he was surprised that they were so close when he was ready to fire. So was the enemy. He kept shooting, three belts, then the barrel burnt up, and it wouldn't shoot. He grabbed his Tommy gun, then he and Mario started the close-in fight.

"Let's get out of here!" He grabbed Mario's arm and sprung out of the trench. Both started running back down the ridge. Joe sensed something, pushed Mario into a crater and dived to the left, rolled, and came up firing. Three soldiers crested the Ridge, not ten feet behind them, their rifles raised. They died in a short burst from his Tommy. Mario, still prone, turned his weapon to the Ridge Crest just as two more soldiers came into his firing lane. The forty-five caliber slugs stopped them cold.

"Getting hairy! Mario! Stop resting! Run!"

"What!" Mario yelled. "I'll show you resting!" He came up firing on more enemy soldiers at the ridge crest as he ran to catch up with Joe.

They passed Captain Hui, who also saw the danger, as he was rallying his men to hold this flank and shouting orders at the same time to start falling back toward the rear exit by the road.

Joe came back with Mario and rejoined their trench with Tony. But Tony wasn't firing. He was leaning against the trench wall like he was aiming his rifle.

"Tony! You okay?" Joe pulled his arm from the trigger and rolled him over. Tony's blood filled helmet rolled off at the motion. "Oh shit! Tony!" He saw the gaping hole in his forehead. "NO! You bastards!" He kept yelling.

Mario saw the charging soldiers in their front, fifty feet away, and started firing.

"Joe! Snap out of it! Grab Tony's backpack! I'm almost out of ammo!"

Joe scrambled, cursing, yelling obscenities at the attacking troops as he dumped Tony's backpack filled with ammo clips and hand grenades on the trench floor. He started pulling pins and throwing grenades in one swift motion, one after another. Boom, boom, boom! Mario kept firing; Joe kept cursing and throwing.

After a few minutes, Mario yelled, "Stop Joe! Stop! There's nobody left."

Red leaned back, looked around.

Foxy was gulping water from his canteen. Tracker was sitting in the trench, head back, eyes closed. More wounded were being brought back to the rear.

Well back, Foxy saw a cluster of men that were laid out on the ground. He watched a soldier with a Red Cross armband move from man to man. Then saw Captain Hui go over to him. Foxy left the trench and went over to the captain.

Watching Foxy, Red turned to Tracker. "Gather all the bazooka rockets you can find, put them in our backpacks. We're leaving in a few minutes. I'll take my bazooka now." Red left to join Foxy.

Captain Hui left the medic and Foxy, walking off with his head down. He started directing his men to evacuate.

"What's going on Foxy?"

Foxy didn't look at Red, just stared at the wounded men.

"These men are all severely wounded and can't be moved. They're going to die. Captain Hui ordered the medic to help them die painlessly. He doesn't want the NKs to play with them. He's giving them extra morphine."

Red stared at the medic as he administered the shots. His stomach turned; he felt like he may vomit. Tears filled his eyes as he walked away, then stopped by a fallen tree and sat on the trunk.

It all came back in one horrible memory. Bobby, you were my closest friend. I'm so sorry I had to kill you. I've never stopped thinking about that terrible day. For your own good, that's what I said. Did you have a chance to live? I don't know? You were severely injured. But these were the Nazi SS. They liked to play with our wounded. You knew that. We saw that. Bobby, I'm so sorry you died. I'm so sorry that I had to do it. I couldn't live with myself if I knew you were going to be tortured.

The memory, like it happened today, overwhelmed him. He then thought about how Captain Hui must feel now, putting to death his soldiers, yes, for a good reason, but still. I wish I could say something to you. Help you ease this horror from your mind. But I can't. Nobody can.

Tracker came over to Foxy, who was still watching the medic. He asked him what was going on. Foxy told him the situation. Tracker nodded sadly then moved closer to the dying men. He gazed at them, seeing but not seeing. He lifted his arms to the clouds, raised his head, and started shouting.

"Brave spirits lie here, oh! *Great One*! They come to you as warriors. Accept them and give them peace. Let their essence wander through your garden of love and forgiveness."

He then went into his Native Tongue, chanted, then spun around and danced and sang, for the dying men.

Captain Hui wasn't far away when he heard the shouting from Tracker. He turned and watched. He saw and felt the passion this American expressed. When it was over, he was stunned. He'd never seen anything like this before. He paused and thought, amazing, this man expressed my feelings in a language I don't understand, in a dance that exuded heartfelt tribute that I feel, and a love for the heroes that these men are. His heavy heart felt lighter. Thank you, mister American. It is my honor to meet you.

When Tracker finished, Foxy came up to him and wrapped his arms around him. "That was beautiful."

Foxy then walked over to Red, saw that he, too, had been affected.

"Red, you okay? We gotta leave! We're ready. It looks like the enemy's ready too." Firing has broken out along the front.

"Yeah, Foxy, just needed to sit a moment. Are we all together?"

"All but Tony, he didn't make it."

"Ah, hell!" He got up, put the bazooka over his shoulder. "Let's go."

Chapter Thirty Three

"What're you doing?"

"Get the hell out of my face, Billy, I've got a mission."

"If you're taking me out of here, then okay. Otherwise, you've got orders to hang around to fly the nationals out."

"In what? We got nothing here, but that piece of shit in the corner that nobody in their right mind would even taxi down the runway. So, get lost!"

"Where're you going in that?"

"I'm picking up gold bullion that President Rhee left near the border, you idiot!"

"Don't call me names, Wild Bill. I'm just trying to do my job."

"Sorry Billy, you're not an idiot, you're just a jackass. Hey, when is that first relief flight due in."

"That's not funny, but better. I got a C-47 due in at 0730 for a quick turnaround."

"Good, I'll need space on that plane for a litter. I'm bringing in a wounded Ranger. Make sure he can get aboard."

Billy's demeanor changed immediately. "Really! Oh! Okay! Sure thing."

Captain Bill Stands finished removing the rear seat from the single-engine L – 5, affectionately named the "*Flying Jeep*." No stranger to the plane, as he flew a few crazy missions in it during the War. His real love, though, was the P-51 fighter. He was a three-time Ace in that plane. First flying bomber support from England, then ground support from France. That's where he first met Colonel Nelson, a Major then. He got his Wild Bill moniker from his fearless, devil may care attitude, both in the air and on the ground. His relationship with Nelson bonded after volunteering for several special operation missions.

As long as he was moving forward in the air, Wild Bill was a happy camper. The faster it could go, as the four hundred per mile per hour Mustang, the more comfortable he was. Even at one hundred miles per hour, the max of the L-5 he was flying, he felt at home.

A simple plane it could land just about anywhere and was specially fitted with a drop-down rear hatch for easy loading. It could haul three hundred pounds of supplies or carry a man on a stretcher.

Billy watched from his jeep as Wild Bill finished up at the rear hatch. When he came around to the cockpit door, Billy yelled out to him, "You're not going up in this now, are you?"

The torrential downpour splashed, made big puddles on the tarmac.

Wild Bill opened the door, turned to him, and yelled back, "Billy! Remember! God is on the side of the righteous! I have nothing to fear!"

He got in, started the engine, and moved off down the runway. Earlier, Wild Bill had taken careful compass headings and timing notes for just this reason. Visibility wasn't much more than a hundred feet in this rain. When the plane lifted off, it turned north and flew low over the airfield to set his course, when he then hit his stopwatch. Captain Stands thought, okay, let the fun begin.

He was flying through the rain clouds at two hundred feet for ten minutes, enough time to clear the northern part of the city. Then he dropped to one hundred feet, where he could see the ground, and followed the only road north up the corridor. The valley was wide here, with no mountains to worry about as long as he followed the road. Going up this funnel, he timed it at maybe twenty minutes, give or take a minute or so.

Korea - The Post – 0630 – (6:30 am)

Shelling begun in earnest around the Post to coincide with the massive attack started earlier at the Ridge. But there was no infantry assault.

Rusty had long ago finished his preparations for leaving the Post as he contemplated in the past. He thought about spending almost a year of his life on this! Was it worth it? Have I done my duty here? Is this the right decision to leave? Thank God they haven't attacked the Post yet. Why?

I shouldn't wonder why, he thought, a bad omen.

"Rusty, men have been moving down the road for a while, mostly wounded being helped." Tully was eyeing the road with his binoculars.

As the shelling intensified, they started exploding all around the Post.

"I see our guys!"

"Thank God!"

"I don't see any more men. I think that's it."

"Did you see all of them?"

"Not sure, maybe I missed one, I only counted five."

"Shit!"

From fort # 2, Snake opened fire with the machine gun into the attacking line of troops coming out of the northeastern woods near where Rusty and Tully observed the Ridge withdrawal. Rusty took a quick look and decided covering the road was not going to be possible. They were facing a severe attack. The minefields no longer posed a serious problem for the advancing soldiers as the constant shelling had ripped the ground apart, detonating or exposing many of the mines.

"Tully, we can't hold! There are too many! We need to get the men moving. I'll be with Larry."

"Go, go, go!" Tully yelled as he ran down the trench line. The men immediately withdrew from their firing positions and started running, all yelling at the man in their front. Tully brought up the rear and was continually looking back. The enemy was fifty yards from the trench. At the tunnel, entrance slowed as the men adjusted their equipment and weapons to climb down the ladder.

Rusty was at the building entrance, watching the road so he could give Larry some timing on when to blow the charges set in the road, so they would discourage the North Korean's from pursuing the retreating ROK forces. Larry was inside the doorway of the building by his detonator switches.

"Get ready, Larry. Got movement now! NK soldiers moving down the road. A tank just came into view. He's past #1 bomb, now on #2, get set! Hit one through four when I signal." Seconds later, he yelled.

"Now!"

Even from four hundred yards away, the sound of the explosions was significant. A massive cloud of smoke enveloped the road.

"No time left for anymore. We gotta go, Larry!"

"Damn! What fun!" Larry said as he exited the building. He thought how much he loved to play with his toys. And now for the finale, the big show. He smiled.

"You got one more, and that should be a blast." Rusty didn't laugh at his joke because he was nervous that it might not go off. That would definitely put a damper on their exit plans.

Last to enter the tunnel, Rusty closed the hatchway and hoped the enemy wouldn't be in the trenches for a few more minutes. The men were all scrambling on hands and knees, moving as fast as the man in front could go. Snake was out first, then Arron, then the rest as they took up positions in the brush. You had to stand up to see the building. That was Rusty's job. Larry uncurled the wires from the tunnel and hooked them to the detonator. Finally, Rusty cleared the exit and knelt next to Larry.

"Ready when you are!" Larry said.

Rusty stood up and watched through his binoculars. He could see swarms of soldiers in the trenches and around the building.

"NOW!"

The concussion shook the trees. The building disintegrated in a large cloud. Much like a volcanic eruption, debris started falling around them.

Tully said. "Jesus! Good thing they didn't have you when they made the Panama Canal. You would have blown the whole damn Isthmus away."

Larry grinned. "You liked it, right?"

Down the road, Red and his men heard the first explosions and realized it was Larry's pet project outside the building, protecting their rear. Now they waited off to the side, knowing the NK's weren't suddenly going to show up. Then the big blast of the Post brought a smile to Red. He knew that they got out.

"They should be along shortly now."

Their friends were coming, and nobody was coming down the road after them.

Korea - The Stream – 0650 – (6:55 am)

The trickle of the creek had become a violent torrent, now a river. It was fifty feet wide and three feet below its banks. Artillery explosions impacted throughout the area but were random and in a wide arc around the Stream positions. The men crossed the old bridge and made it through the ROK lines in search of Manny.

They were directed farther south, where they saw a bulldozer digging new positions off to the east. Soldiers were working feverously to sandbag previously dug trenches. Off to their west, a haggard group of troops was milling about a fire in a shell hole. Covered bodies lay in lines not far away.

"There's Frenchy!" Moose shouted. They could all see the IV pole sticking up from Manny's stretcher in a shell crater, about a hundred feet off the road. Marty started off at a run. Some of the men found nearby holes and lay down. They were exhausted.

Marty examined Manny and changed his IV drip.

Tully went over to Frenchy. "How's he doing?"

"He's hanging in there. We got him this far. I have faith."

"Faith? Frenchy, I love you! You're a north star in the heavens. When the world looks like shit, you see an amber glow. I don't know how you do it, but you make me feel better."

"We French have a gift, monsieur."

"Oh shit, Frenchy! When you start giving me bullshit like that, it usually means things are getting worse."

"Oui!"

"Foxy!" Rusty called out, "See if you can contact our rescue pilot. Nelson said he'd be on frequency 1078. Use *The Post* as your call sign."

Rusty wondered if their pilot got off the runway. It only stopped raining a few minutes ago. Will he find us?

"The Post calling air rescue, do you read?" He kept repeating.

"Rescue calling The Post! Hear you loud and clear! Where are you relative to the Stream? Over."

"Hey, rescue! Glad you showed up! Pick up is about three-quarters of a mile south of the creek/stream/ river, just west of the road."

"Roger that! Just coming up on you! Gonna do a pass to check out the road."

The plane suddenly appeared coming in very low from the south and flew directly over them, then did a slow turn.

"Hey, Post! Can't say I like all the artillery coming in all around you! Road looks okay, though. Can you shake a leg getting my man on board? Not partial to shell bursts!"

"Roger that! We're not too happy about being shelled either. What're you flying?"

218

"Hey, thanks for asking! Name is *Wild Bill*! And I've got one of these super Jet L-5s coming in on you now."

They watched the single engine drop down and land on the road, water splashing with artillery bursts going off nearby.

Tully, Rusty, and the stretcher team ran to the aircraft. Marty was in the lead as he knew this plane and went straight to the rear, and unhooked the loading hatch. He waved the stretcher team over and helped load Manny on board.

The side door opened as Tully and Rusty got near, Wild Bill leaned out. "Nice reception! Next time I want tracers and airbursts."

"I owe you, Wild Bill!" Tully was beside himself. "You ever need anything; you just let me know!"

"Man, I'm not sure who owes what to whom! I haven't had this much fun since my dog ate my little sister."

He slammed the door, revved the engine, did a quick turn, and took off, and climbed into the low clouds, then set his stopwatch. Setting his course, he yelled over his shoulder to his unconscious passenger. "You are much loved, my friend. To have so many men care about you. I don't even know who you are, but I'm now in the circle."

"*Wild Bill*! He sure knows some strange guys!"

"Who?" Rusty asked Tully.

"Colonel Nelson."

"Let's get some est."

Chapter Thirty Four

Flushed, General Walker stormed into the conference room.

"Colonel! How come I have to learn about a major event in Korea from my Air Force liaison staff sergeant?"

"Guess you're talking about the order to evacuate American nationals."

"Right! Aren't you supposed to keep me informed? That's important, isn't it?"

"Yes, sir. Unfortunately, I just learned about it myself about an hour ago from a corporal at the Kimipo Airport, and I've been dealing with a crisis since."

"You mean you didn't get anything from General MacArthur's headquarters?"

"No, sir! I called General Almond's office as soon as I heard, spoke with his night duty officer. He checked the order distribution list, and Eighth Army wasn't on it. He thought it must've been an oversight."

"Oversight! Incompetence! That's what it is! I apologize for barging in here like this. I couldn't imagine you'd be remiss, but I'm pissed!"

"General, I'm glad you didn't shoot first and asked questions later. I'm not sure what I would've said to General Almond if I'd gotten to speak with him. I was pretty upset with him myself, still am."

"Jim, this has become a political football. I think you pointed that out a while ago. What's that famous saying, *don't get caught offsides*. You be careful. General MacArthur is a quintessential politician; his staff mirrors his ego. It's his show, and he's going to run it his way."

"That reminds me, General. If you don't mind me being so bold, but you had a certain unflattering look on your face at our briefing yesterday, when General Almond commented on the South Korean soldiers. I was curious about that."

"You don't achieve rank without a strong political streak. Besides brains Jim, you've got that gift as well."

"It's an innocent question, General. What do you mean?"

"It's not so innocent, and you know it. If you don't, you should, being the all-knowing G-2 of my Army. So in case you're not so political, which is why I admire you, I reacted that way because I have grave misgivings about our General Almond."

"That's serious stuff, General. May I ask why?"

"You may because you need to know the character of this man so you might protect yourself in the future. He's a Virginia blueblood, went to VMI along with General Marshall, and is his friend. He got promoted over many more worthy candidates to become general and placed in command of the very first all-black division in 1942, the 92nd Infantry Division, I believe. They were part of the Italian Campaign in 1944- 45. The division did so poorly that they were withdrawn from the front line and regulated to rear support missions. He was famous for castigating all black soldiers for their lack of inherent fighting ability and spirit."

"Almond blamed his men for his division's failings?"

"That's the kind of person we're dealing with."

"I didn't know."

"The word is pretty much out to most, except, of course, General MacArthur."

Frustrated, the General shook his head. "Alright! Enough about this! What's the current situation?"

"It is, as we suspected it would be. Our military situation is almost non-existent! Beyond the evacuation orders, our government has not taken an official position yet. We are doing something, though; General MacArthur has authorized all Services to "render assistance" to the South Korean military forces regarding supplies and ammunition. Also, our Ambassador to the United Nations submitted a request for an emergency meeting of the Security Council to address the Korean situation."

"That's got to be our President pushing that! Glad I voted for him! Go, Harry! Didn't mean to interrupt Colonel, but I suspect the rest of your brief will put me in a bad mood."

"I believe it will, General. I just spoke with my G2 Korean counterpart, Colonel Pyong Oh, and he's at his wit's end. The ROK Army Headquarters staff is losing control; communications are breaking down. North Korean forces have broken out on three of their four main attack routes. They lost complete contact with their east coast forces, with the enemy seemingly unopposed heading down the East Coast Highway. The west coast is not faring well, either. I'm trying to help rescue that ROK regiment trapped on the Ongjin Peninsula. I told Colonel Oh that we have two LST's in a nearby port that could lend support, we're both working on that."

"MacArthur's got to approve it."

"That's what I told him. I suggested he have his Chief of Staff call General Almond."

"Good. Need I remind you to be very, very careful about stepping on toes connected to big feet."

Colonel Nelson acknowledged and continued his briefing.

Korea - South of the Stream – 0900 – (9:00 am)

Sun broke through the clouds.

Increased intensity of nearby explosions roused the men. Most had slept in shell craters or a ditch, but they kept close. As they stirred, Tully eyed soldiers filling sandbags nearby. Yup! He thought, no sandbags gonna stop what's coming down this road! We've got to move.

Tully found Rusty

"Glad we got an hour or two of rest. I know everybody's worn out, but it looks like it's time to get someplace else because hell and brimstone are about to descend right here."

Rusty looked at Tully, "You're a gifted soothsayer."

"You're a good judge of men."

Both were silent.

"Damn shame about Tony. Were you close?"

"Not sure he was close to any of the men, Rusty. But no, we weren't close. Tony came to the squad about two years ago. He was in Italy with the 3rd Rangers. Kept to himself and was a well-spoken quiet guy that liked to read. He was excellent on that mortar, wasn't he?"

"Red tells me he took out a lot of men, too, at the Ridge. Not a bad way to go if it's your time, I'd say. And talking about time, I'd mean that you're right about moving. Let's get the men away from here now. We'll head south towards our depot, a few miles or so. Get Red. He knows the way. First, I need to talk to Captain Hui."

Nearby, he saw Foxy sitting in a crater with Tracker and a group of ROK soldiers. "Foxy! I need you with me!"

They walked toward the new positions Captain Hui and his men had taken. Rusty scanned down the defensive line and thought how few of the South Korean men had survived.

"I'm glad you made it, Foxy." Said Rusty. "Must have been rough on the Ridge."

223

"Yeah, it was." He paused, "On many levels. Joe saved our asses, thought you should know. He and Mario stopped the NK's from flanking us and rolling up our line. They both should get medals."

"We should all get medals, Foxy. Hey, what were you up to with those ROK guys?"

"Their bazooka teams at the Ridge didn't make it. The one we left here got destroyed in the shelling. But we recovered one and thought that we didn't need two with what we're about to set out to do, so I was instructing these men on how to operate the weapon."

"Did you check this with Red?"

"No. He seems a little strange right now, distant, but I didn't think he'd object. Do you?"

"Just being cautious, but no, I agree with you. We have enough to carry. One is enough. What happened to Red?"

"Not really sure, he was affected by a horrible decision that had to be made at the Ridge." He told Rusty about the wounded men that couldn't be evacuated.

They found a despondent Captain Hui, sitting alone, head in his hands. Rusty and Foxy sat beside him and began telling the captain how heroic he and his men fought at the Ridge.

He didn't respond, sad and haggard looking but not quite in shock, but close.

A big truck convoy started pulling up off the road just behind them, stopping for a minute while troops piled out, then turning and pulling out, heading back south down the road.

A jeep pulled up. An officer got out and started shouting in Korean to another officer closer to the men coming off the trucks. He then came over to them, must have noticed the Americans arm patch because he stopped and saluted. Curiously, Rusty, and Foxy both stood up and saluted back. Captain Hui didn't budge, kept his head down. Foxy spoke first in Korean.

"Colonel, my name is Foxy, and this is my fellow officer, Colonel Rusty. He is in disguise as I. We're both American officers who have been fighting since the invasion. Our brave friend here is Captain Hui, who defended the Ridge for the last two days. Please pardon him, as he suffers from exhaustion and the realization that he lost most of his command."

The colonel's expression changed. He nodded in acknowledgment.

"My name is Colonel Yup. I've met you before, Colonel Foxy, a few days ago. I was told you might be here. I'm glad you survived. Your help meant a great deal. I am in your debt."

"Yes, of course, Colonel, please forgive my forgetfulness. No need to thank me for doing my duty."

Captain Hui came out of his fog after listening to the exchange. He looked up and recognized his colonel; he slowing stood and saluted. He apologized for his lack of manners and insisted that no disrespect was intended.

The colonel walked up to the captain, put his hand on his shoulder, and told him he was the best officer in his regiment and how proud he was of his stand against an overwhelming enemy. A crowd of soldiers gathered around the small group.

Standing in front of Captain Hui, Colonel Yup got ramrod straight. He spoke.

"I wish to bestow on you a special honor, Captain Hui!"

Soldiers with cameras came alongside and got into a kneeling position. Cameras started clicking.

The Colonel continued.

"By order of our President Syngman Rhee and our General of the Army, I at this moment award you our country's highest award, *The Order of Military Merit, First Class*, for your bravery and leadership at the Battle of the Ridge." He drew the medal from his pocket and placed it around Captain Hui's neck. The medal's appearance was similar to the *United States Medal of Honor* and had the same meaning.

He then saluted Captain Hui.

Curious, some of Captain Hui's soldiers had gathered around wondering what was going on and witnessed the event. They started clapping and cheering.

Foxy was speechless. Rusty knew something extraordinary had just happened. He saw the beaming face of Captain Hui. A far cry from the look he had seen a few minutes before. The men started shaking hands and congratulating the captain.

Trucks continued coming, dropping off more soldiers. Shells randomly exploded nearby.

Rusty leaned close to Foxy. "We have to go. It's time we get out of here before all hell breaks loose." Foxy nodded.

"Colonel, we must catch up with our men. I hope to see you again soon. Best of luck here."

"Where are you going? One of my trucks can take you and your men."

"Appreciate the offer, but no."

Embracing Captain Hui, they bid him good-by. Mutual respect and fondness were palpable.

Starting to leave, Foxy heard the Colonel ask, "You didn't tell me where you're going?"

"Sorry, Colonel, I can't. All I can say is that we're here to help your country. That's our mission."

As they walked to the road, Foxy translated the entire event to Rusty.

"Colonel Foxy! I guess you expect me to start saluting you from now on?"

"Colonel Rusty! I'll salute you if you salute me." They laughed.

Red and the men had taken their time, heading south away from the oncoming onslaught, elevated everyone's mood. The sun was out and the morning was pleasant, spring-like with a slight breeze. The constant bombardment to their rear and the passing trucks was the only disconnect to their enjoyment of nature.

"Mario, you smell like a dead coyote?" Snake said from behind him.

"And here I thought we were friends. I've not said anything about your aroma, which by the way, I would closely associate with a two-day old salmon fish kill."

"I think you're picking up Tracker's scent behind me because I have a hard time breathing the noxious fumes he's putting off."

"You white people stink when you wash. You have no respect. I'm trying not to vomit walking behind all of you palefaces."

Joe was behind Tracker.

"I was ten years old when Lou and I stole food from the Fulton Fish Market in Manhattan and got caught. We were thrown into a vat of fish cuttings and refuse. That smell has always eluded description. That is, up till now! God, we all stink!"

Arron was alert to the whole repartee, so when he finally got his turn, he was loud.

"Not so fast, you New York sharecropper! Not all are painted with the same smell! I say Northerners stink more than Southerners and people west of the Mississippi are just weird with a musky something odor, but, no offense, Mario, I can smell you from here, and you really do stink."

Red led the men off the road and back to the mountain ridge area where there was a nearby hidden jeep and arms depot.

"All those who stink the most take a break downwind of us. In case you don't know who you are, I'll give you a hint. If you are covered with dried blood and participated in some recent assassin activities, you would be a prime candidate. All others who just smell badly are free to mingle."

"Hey, Mario." Snake called over. "I think Red's talking about us, let's grab a seat over there." He pointed to a small clearing under a tree, not far away.

"Did you take C rations with you? I brought some extras."

"I got a few, too. Man, I didn't realize how hungry I was." They both went through their backpacks to find food.

"Red told me you and Joe did some crazy stuff over at the Ridge, sounds like things got kind of interesting?"

"Nah, just typical Ranger stuff. Joe, on the other hand, well, I'm not sure how to describe what he did. He acted like a mixture of a Ranger blended with a Marine, with a dash of Houdini and a splash of clairvoyant."

"Mario! That really paints a clear picture of events. Do you have a degree in Abstract Mountain Climbing to go along with your degree in Creative Speaking?"

"When you got talent, Snake, you just got to show it off. I can sense some jealously. Please don't create a rift in our relationship. We have a lot of assassin stuff to look forward too, so don't spoil things."

Snake shook his head with a smile on his face. "You are one weird and crazy hombre."

Frenchy sat with Joe nearby. Both had pulled out C-ration cans. Joe leaned over to Frenchy and tried to glimpse at his can to see what he had but couldn't.

"Hey, Frenchy! I got a can with no label. I hate surprises. I'd like to know if this is safe. Do you have a label?"

"Safe! It's US of A issued, of course, it must be good, but let me see." he raised the can, elaborated his inspection and pronounced, "Lobster Bouillabaisse, with extra shrimp. Ah, my favorite. Let's see, oh yes, good, it's freshly prepared too, March 1945."

"Would you trade me your delicacy for a no-name can after you tell me what you actually have?"

"No monsieur, we French are sensitive about knowing what we eat."

"Frenchy, you're an upstate New Yorker. Don't be an asshole. You know that you wouldn't last five minutes on Flatbush Ave, so don't screw with me."

"That's right, Joe. Three minutes is all it would take me to get the most beautiful girl on Flatbush Ave and take her away."

"Right! So what do you have? I'm game!"

Rusty and Foxy caught up with the men, sat down near Tully.

"Tully, has Red done a count on our rocket ammo for the bazooka?"

"Yeah, Rusty, grab a C ration and relax! You've got good people all around you."

"Of course, I know that. It's just me."

"Right!" Tully nodded, "So, just being my curious self, what are we going to do now?"

Foxy looked at Rusty, and both smiled.

Rusty then said, "Let's just say that your Ranger self will be thrilled."

Chapter Thirty Five

Monday - June 26, 1950,   Day Two of the Invasion
Japan - Eighth Army Headquarters – Colonel Nelson's
Conference War Room - 0900 – (9:00 am)

"Colonel Nelson! Colonel Harris on line two."

"Did you get those bazookas ordered?"

"You're funny this morning, Jim. I'll be just as funny. Eighth Army got marching orders. Well, two squads of the Army, anyway. They'll be on the ten AM flight to Kimipo and be deployed for base security during the evacuation. I wanted to give you a heads up."

"Son of a bitch! Two whole squads!"

"Oh, one of the squads is being diverted to Suwon Airport, twenty miles south of Seoul."

"What's at Suwon?"

"Not much, but it's a good backup in case Seoul folds quickly, and it's away from the main action. It also has a few warehouses stocked with surplus stuff that Eighth Army didn't want to ship back to Japan when we pulled out when the country became independent."

"Any weapons or ammo?"

"Probably both, but I have no information on what's there. I doubt that anybody knows."

"Give me a call if anything useful turns up."

"Sure thing, and oh, you have a nice day."

Very funny, he thought as he hung up. Damn it! I still haven't heard anything from Wild Bill or Billy. He picked up the Red *Kimipo Airport* phone. Many rings but no answer.

"Sergeant Crowley," Colonel Nelson yelled over to the duty station aid, who immediately rose and came over.

"Sir!"

"I want you to find out if a wounded Ranger was evacuated from Seoul this morning. Probably on a flight, check that first. If he was on a flight, I want to know where he is now. His name is Manny. I don't know his last name."

"Yes, sir! ASAP! Shouldn't be hard to trace."

Nelson got up and walked around the table and stopped. His men were answering phones and moving to the maps on the wall, active. Yes, he thought, very intense. Trying to figure out how to save thousands. That's what I do best, isn't it? It's what we do best, goddamn it!

His mind went back and forth. What's going on, Colonel? Not sure, I've done this job for a long time, so what's up?

Pretty easy to send Platoon A into a kill zone, isn't it? A little different when you have to send Manny or your entire handpicked Korean team through death's door? But to save one! You know, you can hang your hat on that, right?

The red phone light blinked and buzzed, brought his mind into focus. He lifted the Korean *G2* phone, "Colonel Nelson."

"Colonel Oh, here. Much thanks to the LST's rescue of my regiment! That's happening as we speak. I want to return the favor. I just seized two tramp cargo vessels in Inchon Harbor before they got underway. You could use them to evacuate your nationals. The airport is not going to be able to handle them-too many people. I'll hold the ships for two hours. I have many trucks just returned from the front. I have no more troops to send North. Think I have a few hours before I'll need them to go South. They are yours if you want them. They could be at the airport in thirty minutes. Let me know!"

"Wait! You have any word from the Ridge or my Post men?"

"I have no details, only that our unit abandoned the position after a fierce battle. I've got to go."

Damn! He thought. Well, **abandoned** is not overrun. Good news about the ships. The harbor was only a twenty-minute drive from the airport. They could get most, if not all of the Nationals out in a few hours. Not the days needed if they stuck to planes. They don't have days.

He picked up the white phone and actually spoke directly to General Almond. He approved of the change. Nelson called Colonel Oh back, gave him the okay to release his trucks, then called Kimipo airport, got an Air Force captain who explained that he was in charge of the evacuation and air coordination. Colonel Nelson told him about the new situation, alerting him to the trucks and updated his orders.

"Colonel, Colonel!" Nelson raised his head to see who was yelling at him. Sergeant Crowley hurried toward him.

"What is it, Crowley?"

"Sir, I found Manny!"

Chapter Thirty Six

Tracker led the team away from the hidden depot, without digging up any of its supplies. They carried plenty, figured they could come back for more. Through dense brush, he led them into an old dried up stream bed that led into the foothills. The path was rocky that headed up through rough terrain, then quickly ascended to rising foothills into the mountains, and kept rising. Sometimes, it angled at an absurd 45-degrees. No one spoke, except for the occasional warning of this edge or that loose rock.

Rusty followed Tracker up the path. "Are we almost there?"

"Do you see the top of the mountain?"

"I don't see anything but this tiny ledge and a huge drop into nowhere."

"I should have put Xs on the paths. Hope this is the right trail."

Rusty stopped and leaned against the mountain wall for stability and grunted.

"You make me laugh again on this tightrope of a trail, and I'll tie you to a pole and whip you just like they did to sassy Injuns in the Old West!"

"Spoken like a true paleface! I'm almost ashamed of you, talking to me like you think you could catch me. Lucky you that I long ago made my peace with white people."

"Is this some revenge you kept in your head when I ordered the stocking of this Peak?"

"There's that. Two days of lugging all that shit up here. It's not easily forgotten."

"What the hell is holding us up?" Yelled Red, four men behind.

Rusty yelled back. "Oh shit! Nothing! I'm just afraid of heights!"

"Damn liar! You ain't afraid of anything. Stop screwing around!" Red yelled back.

"Not far." Tracker said back to Rusty.

They rounded a pinnacle and entered into a flat area. A waterfall cascaded off to the right into a pool that overflowed to their left. It was a small but beautiful area, like at a fancy resort. A sloping shoulder close to the mountain wall left room for the men to rest. After climbing for two hours, heavily burdened, soaked in sweat, the men were beat.

Red scanned the cliff above to the top. "Rusty, there's no sign that Captain Jordan or anyone else is on The Peak."

"How can you tell? I can't see anything."

"That's the point. We should see a lookout."

Joe was the first in the pool.

An hour later, the mostly naked men were lying on the rocks, strewn clothes lying about, drying in the sun. Distant rumbling reminded them of the nearby danger. Tully sprawled out next to Rusty and Red.

"I wish we could just stay here all day." He said to the blue sky.

"Glad we're not going to take a vote," Red said. "We've got too much to do."

"How much further, Rusty?"

"I have no idea, Tully. Only Red, Tracker, Larry, and Arron have been here. What say, Red? I'm pretty curious myself?"

"Something like three hundred feet or so, straight up. It's a climb but doable. Tracker! How many times did we go to the top?"

Tracker frowned. "More times than I'd like to think about right now. Good thing it was winter. But let's see, me and Arron made the first trip on our recon mission. Then you came on the inspection mission. Then we spent two days hauling the stuff up to this pool.

That was what, four trips each day? Then a day of hoisting to the top. So eleven trips to here and three to the top."

Tully was puzzled, "Hoisting?"

Getting dressed nearby, Arron heard the conversation. "Our mechanical genius came to our rescue again! Red-figured that if we were going to stock our mountaintop hideout, it wasn't going to happen by climbing it up. You'll soon see what I mean. Two free hands are needed, and balance is necessary. Too much weight on your back isn't good."

"Arron! Are you writing a book? Jesus, do you go on." Red said to Tully. "It was simple, really; I rigged up a line to a pulley. We had an earth moving canvas tarp, so it became the holder. Construction at the Post was almost finished, so I borrowed all this stuff. Nature cooperated by having a fallen tree in a convenient spot. No genius here, just luck and good timing."

"I hate it when you spoil my fun, Red! I was starting to think that maybe you had a relative from Virginia or maybe even farther South."

"Shut up, Arron!" Red got up. "Men! When we get to the top, do not go into any of the caves. Larry has to disarm all of his bobby traps first. Stay back under the trees."

Rusty said, "Okay, let's get to our vacation home! The women are waiting for us!"

The men started moving, getting dressed, and picking up their gear. Frenchy whispered in Arron's ear, "You never mentioned any women! Are there enough?"

Arron slapped his leg and belly laughed.

Chapter Thirty Seven

The ball was starting to roll a little faster now as the gravity of the situation began to impact old notions that were never valid. The invasion was moving so swiftly that nobody at MacArthur's headquarters in Japan realized that Seoul was going to fall quickly.

Colonel Nelson was sitting in his chair, on the phone constantly, for many hours, urging for more action, trying to get better information, etc. He was frustrated. Despite his best efforts at warning about the impending collapse of the South Korean Army, the United States Political and Military response remained timid. Get going, he said to himself. You never give up! Keep trying!

Concluding a long time ago that it was sometimes best to go down a few rungs in the command structure to actually get something done, he had some ideas. Though this had limitations and some danger, he tried to keep in mind what General Walker advised him about trying to stay away from getting under *big feet*.

Colonel Nelson formed an idea around the orders that MacArthur's command had issued to assist in the evacuation of US Nationals. Nelson thought that it was a pretty interpretative order and one which he might have some room to maneuver.

He called out, "Get me Air Force Colonel Dietrich at Itazuke Airbase!"

USAF Colonel Dan Dietrich commanded the 45th Tactical Reconnaissance Squadron based in the city of Fukuoka, Japan, which was a hundred miles directly across from the tip of Korea and the city they call Pusan.

"Line three, sir!" The sergeant said. "Colonel Dietrich!"

Nelson picked up. "Dan! I need a favor."

"What? No, how's the family, how are you?"

"Glad you know me, so you know this is important. It's about these new orders we received from MacArthur's HQ, Japan. I quote, "assisting US Nationals in their evacuation." I need to know if you consider gathering aerial intelligence about enemy troop dispositions part of this directive."

"Skirting an order can be detrimental to one's career, but I'm game! God knows you saved my ass more than once. I owe you. Let's have it. Tell me what you want?"

"I need an aerial recon mission from Seoul north up the Cheorwon Valley to the border and then south down the Uijongbu Corridor to Seoul. I need a fast turnaround on the photo analysis of the same, like an hour or so."

"You always want the impossible. Why is this so important?"

"Good question. It's secret. I have men there that shouldn't be there, and their lives are in danger. That's all I can say. I could go on and tell you how our Nationals are threatened by the rapid advance of the North Koreans, and how their evacuation is in danger, etc. But I won't, but that's true."

"Still pretty lame Jim, can't use your *secret mission,* and there are no Nationals in these valleys."

"Well, what if there *were* Nationals and they're trying to escape. Say, like two Catholic missionary schools, one in each Valley. That should justify a recon mission, right?"

Silent for a moment, Dan said, "You're a clever bastard. I like it!" He paused again, "but I think they should be Episcopalian schools."

Nelson laughed. "I agree, not many Catholics above us! Is that enough cover for you?"

"I don't need much cover, Jim. You're making me go way beyond being creative, so this will have to do if anybody gets technical. I'm already flying recon over the coasts and have fully prepared all of my photo surveillance aircraft for immediate duty. My Photo Analysis team is already digesting the latest film and can respond to new data very quickly. I'm ready."

"How ready?"

"Pushy as ever, Colonel! I've got a P-51 on the runway with another three behind it. I've been looking at these valleys while we speak. Cheorwon is about seventy miles long from Seoul and the other only fifty or so. The border area is pretty iffy up there. I don't want any over the border stuff now. I couldn't even begin to justify that. So I think I'll send a plane to each valley. They'll go slow and low and be out of there in fifteen minutes. What area is most important to you?"

"My immediate concern is from the town of Cheorwon south ten miles."

"I'll process that first, probably take three or four hours. They'll lift off in fifteen minutes. Does that work for you?"

"Works for me! Thanks! I'm now the new debtor."

"You damn sure are!"

Reconnaissance Flight heading to Korea
Over the Korea Strait at 10,000 feet – 1200 – (12:00 noon)

"Hey, Nosepicker! What if we meet up with bandits?" Captain Hurst heard on his radio.

"Stop being stupid. If you can't out run'em, do something crazy. Pisser! Stay off the radio and follow our plan."

The plan was that when they reached Seoul, they would drop to three thousand feet, split up and get the cameras going. Nosepicker would go right up the Cheorwon Valley and Pisser straight up the Uijongbu Corridor.

Captain Hurst's P-51 had a large painted human nose on its front, with bullets entering its nostrils and flames coming out its side. Hence his call sign. He often boasted *I could shoot the snot out of a Jap at five hundred feet* and had regularly backed up that claim over Okinawa and Japan in the War.

"Here we go! No funny business Pisser, see you back at the barn."

Korea - The Peak –1230 – (12:30 pm)

Ben was on guard duty at the eastern edge of the Peak. What a beautiful view, he thought. Despite the rumbling of war in the distance, he felt relaxed and safe. Rarely had he ever felt safe in his life. People were always dying around him. It started early. Reflecting on his past, Jesus, I must be cursed! First, my oldest brother, then Al, my best bud. But I got that shithead who killed you. For what? Why did they hate us so much? Why do I hate so much? He answered his question, remembering how he killed that bastard. It felt good to get revenge, yeah it did. Nobody was going to dick with my family or me again, ever! But they have, haven't they? He started listing his buddies, who had died by his side. It was a long list. He pondered this then straightened up. My friend Manny has a chance. I hope he makes it. Maybe my curse is over.

A far off sound brought him back. Damn, if that doesn't sound real familiar, got to be! That humming sound of its engine is so distinctive. He got his binoculars up, trying to anticipate the progress of noise as it grew coming up the valley. Yes, there you are! The P-51 came up and flew close by, just a little higher, going straight up the valley. Wow, he thought, we're not alone out here.

What the hell? A flaming nose?

Dennis Kennelly

Japan - Itazuke Airbase - 45th Tactical Reconnaissance
Squadron Photo Analysis Lab
1400 - (2:00 pm)

The room was crazy with activity as the film from the two recon flights was processed. Each plane carried three cameras, and each film had to be separately processed. Fortunately, the camera run was only sixty miles or so and held a total of fifteen minutes of film. At two frames per second, there were only a total of eighteen hundred frames per camera that had to be reviewed.

Colonel Dietrich had given the analysis team their priorities. First processed and analyzed was the film from the center camera on the Cheorwon Valley flight. The rest would follow after a full detailed study of this film. At three thousand feet, the pictures would be sharp and cover the three-mile width of the valley from mountain to mountain.

After giving a time frame for a preliminary analysis, he called Colonel Nelson with the update.

An hour later, a tech sergeant from his team interrupted him. "Sir! We found something I think you should know about."

He followed his sergeant to the lab; soldiers were reviewing a wall projection of one of the frames.

"Where is that?"

"Colonel, that's five miles south of the town of Cheorwon. It is the only bridge in the whole valley, and as you can see, it's functional." A long line of vehicles stretched north, waiting to cross the bridge and an equally long line extended south of the bridge.

"Is that a railroad bridge to the east?"

"Yes, sir. It, too, is undamaged."

"This is imperative! Good job! When will you be finished?"

"Another hour or so, Colonel."

He called Colonel Nelson.

240

Chapter Thirty Eight

Monday - June 26, 1950,   Day Two of the Invasion
Korea - The Peak –Same Time - 1400 – (2:00 pm)

"That's what I'm ready for." Red looked at Rusty, sleeping under a shaded tree.

"We deserve a break too! Let's grab a tree." Tully walked over to a nearby tree, sat against it. Red followed.

The mountain air was fresh with a slight wind, and at an elevation of about three thousand feet, the temperature was a mild seventy-eight degrees. Tree canopy on the peak was significant, strangely growing out of ancient rock formations and occasionally, at odd angles. The flat area that was their camp was expansive, with rock formations on either side going up at steep angles, for maybe another three hundred feet higher.

A series of caves lined the two sides. On the north side were three caves, one large, four feet wide by five feet high by fifteen feet deep. The smaller caves were both similar, two feet wide by three feet high, and about ten feet deep.

The south side had one cave, and it was big. The entrance was small, three feet high by three feet wide, but after entering, it grew bigger, enormous. Within a few feet, a six-foot person could stand and see a twenty-foot wide and deep chamber that went well back into the shadows.

After a while, Tully and Red went over to sit with Ben, discuss the implications of the American fighter's presence in the valley. Coming to no conclusions about the plane, they moved to a corner, away from Ben, and focused their attention to the majestic landscape and beautiful mountains to the east.

"This feels like Bavaria."

"I know. It brings back memories."

"Me too. I had the privilege of liberating a mountain top retreat of one of the Nazi big shots. It was empty except for one of the caretakers. Views like this, but much better, with a large lake at the base. And boy, did this guy like his wine! He had one beautiful wine cellar. Must have been a thousand bottles in there, all stacked sideways in wrought iron racks, French Champagne and wine of every sort. We had a two-day bender that was hard to describe. A good memory."

"Mine wasn't. I'm still not sure how I feel about it."

"What happened?"

"I've never told anybody about this." He reflected a moment. "I think you'll understand." Red looked at Tully with a pained expression on his face. Tully just nodded his approval to go on.

"I led a team into one of those Nazi chalets on a mountain, in Bavaria, probably near where you were. We were to arrest an SS General. Looking out from his grounds were mountains and a green valley, just like here, beautiful." He went on.

"My unit had just freed a death camp the week before. He was its commander. That's why we had orders to arrest him."

Tully related. "I've been through one of those camps."

"Anyway, this asshole came out of his chateau in full uniform, medals hanging on his chest, holstered sidearm, boots spit-polished and arrogant, very arrogant. He wanted to surrender with class, all formal, a pompous little shit. *I'm a General*! He said in English. *What are you, a sergeant? I will only surrender to your general. You pion! Do you know who I am? Get your general now!*

"Don't know what happened to me. I'd never seen atrocities like that before in that camp—piles of bodies, women, and children. That picture flashed through my mind, and I lost it. I pulled out my .45 and shot him in the head."

Tully nodded and paused. "I think we were talking about how beautiful it is here."

Korea - The Peak – 1530 – (3:30 pm)

"Hey, boss!" Larry stood behind Rusty. "I've got the VLF hooked up; we can send and receive."

Rusty cleared his head. "You confirmed contact?"

"Big Roger on that. Got preliminary code feedback from Colonel Nelson's command."

"I'm glad you didn't fall off the mountain hooking it up."

"Mario climbed the three hundred feet to the top to place the cone. Not sure I could've done it alone. The weather is so clear we spotted the tower without using our map reference coordinates."

"You're a genius, Larry. Who's on the electric wheel?"

"Moose."

Rusty got up, as did Red. Saw Tully and waved him over.

"Red! Find Joe."

They entered the main cave and positioned themselves around the VLF set up. It was much like a short wave radio, reconfigured into a Morse code send/receive apparatus.

This VLF was a brand new technology, requiring little electrical energy to transmit a message on a frequency so low, that it could travel almost forever. Its only drawback was that it would only move in a straight line, somewhat like a rifle shot. The signal did expand slightly over distance so that exact precision wasn't necessary for connection to a sending or receiving cone.

"Joe, I want you to transcribe the incoming code. Larry, you focus on sending." Rusty was all business now.

"Larry, start the call. State our position and request an update."

Larry's finger hit the Morse code clicker. Moose was cranking away on the generator wheel nearby. The clicking and clanking sound eerily reverberated in the cave as the men stood by. Larry finished entering the verification codes, then completed the message.

"Standby" Joe responded to the message received.

In a few minutes, the click sound started.

Joe was recording the message and speaking at the same time.

"Manny critical, but stable. Is Captain Jordan there? Keep to our plan. Air Recon active. Air support, maybe. Cheorwon captured. Road and railway bridges south of town intact. Destroy, if able. NK in Valley to twenty miles north of Seoul. Must delay NK at all costs. N."

"Larry, send this."

Larry coded as Rusty spoke. "Grateful for Manny. Nobody with us. Will review bridges for op tonight. Need update on troop disposition in area. Need radio frequency for air support. Will contact later before op. R

A few minutes passed. Then the clicking started. Joe translated. "Use 1096 for air. Use code Post to Unite. Later for aerial update. N."

"Can I stop now?" Moose was still cranking on the generator.

Larry looked over, "Yeah, buddy; you can stop."

"Let's check the maps. Red, you ever been to Cheorwon?"

"Pretty sure only Foxy and Tracker have ever been over there."

Red, Tully, and Rusty moved outside, gathered around a shaded rock outcropping, and brought out their map book. Rusty looked around but couldn't see either man, so he yelled, "Foxy! Tracker! I need you here. Now!"

Back in the cave, Larry was checking the small VLF box, making sure that the small battery was on so that a red light would blink if there were incoming traffic.

"We never were able to get into the details of this VLF stuff, but I have to tell you, it's impressive. It seems to me it's like the old telegraph hardwire system but wireless. Is that kind of it?"

"You're smart, Joe. That's the simplest description I've heard about this system. This transmission technology has leaped into another orbit. It's still in its infancy, but it's already changing everything, and we have it."

Not comprehending, Joe shook his head. "You sent an uncoded universal Morse code message, to where, exactly? Why can't the enemy pick this up? Sorry for being dense, but I need to know! I am your backup, right?" Joe had an edge to his voice.

Larry leaned back, got settled against the cave wall, and looked at Joe, maybe for the first time.

"I'm on your side, Joe. No need to get testy with me. If you haven't noticed, I'm a technical person. Demolition and electronics are my love. But I can fight and have. I will tell you everything you need to know. You and I just haven't had the time, have we? Until now."

"Sorry, Larry. Didn't mean to infer anything. I just realized our survival depends on this new stuff. If you go down, it's on me, and I don't know shit about it."

Larry showed Joe the map of the hidden transmission towers spaced every thirty miles or so down the central mountain range, how to nuance a connection, and all the rest. They talked for another hour.

Foxy was taking a leak off the cliff edge when he heard Rusty's call. Damn! He thought I've never pissed into a beautiful valley before. Better be important! He finished and found the men.

"We need your help, Foxy. When we first started setting up at the Post, you and Tracker reconnoitered the town of Cheorwon and this valley." Rusty pointed off to the northeastern edge of the Peak. "Cheorwon was captured, and its bridges south are open to the valley, all the way to Seoul. We need to do something about that."

"Come look at the aerial map book. What do you remember about your trip?"

"Rusty, hold up a minute. That must have been six or seven months ago. Besides, I was driving, and Tracker was doing most of the looking. I didn't see much because I was trying to avoid getting the jeep stuck in one of the many holes on that road. Let me get Tracker." He left to check the caves.

Tully was looking at the map and then turned to Rusty. "Didn't you tell me that there was another Post set up north of this town? Is that where this Captain Jordan was that Colonel Nelson asked about?"

"Yeah. There was a Post north of the town, closer to the border, but it wasn't like ours. It was abandoned the day before the invasion. At least that was the plan. They were supposed to move here."

"Didn't they have an escape plan like you?"

"Yes, all four Posts had withdrawal and fight plans. That's what's disturbing. Since the Peak straddles both valleys, this position is called Peak One, and it was to be our first big fallback position for both our Post and the one north of Cheorwon, Captan Jordan's post. We have more locations farther south. Colonel Nelson told me they were about to move south of the town to wait and see what happens. He's not heard from them since."

With a sad awareness on his face, Tully nodded

Foxy arrived with Tracker grudgingly, bringing up the rear. He looked disheveled like he'd been on an all-night bender.

"What the hell's wrong with you?" Rusty had never seen Tracker like this.

Tracker sort of looked around, not sure where he was. "Wow, those mushrooms are better than my homegrown stuff."

Arron walked over with a cup of coffee and handed it to Tracker, then looked at Rusty, "He told me he found some mushrooms in the back of the cave and thought he knew what they were. I told him they were not the kind you eat; he just laughed."

"I can't believe this." Rusty was beside himself.

Tully started to laugh, then Red, then Foxy, finally Rusty.

"You made coffee?"

"You're graced with Southern hospitality. Please follow me to the café."

It was evident that Tracker was not ready to shed any light on much of anything, and the men hadn't had anything hot in…most couldn't remember when. They followed Arron into the cave and sat on boxes or leaned against the wall. In front of them, were a series of four steel helmets, each over a small fire with a branch stick set up to support a hanging piece of cloth in each pot of heated water. The smell of coffee was intense and wetted the men's taste buds. Arron had raided the stacked backpacks for the tin cups and had set them out. The men grabbed them and dunked them into the brewing helmets.

Tully took a long swig. "Heavenly!"

"Ah, you have chosen the "Coffee of the Savannahs." Arron made an expansive gesture. "My favorite."

Rusty was first to notice that Arron was in bare feet. He took a closer look at the cloth hanging in the steel pot and recognized the color "You are a real chef!" He raised his cup to Arron and burst out laughing.

Later, after Tracker had his fill of coffee and regained his mind, the men regrouped outside around the maps.

"I remember our trip well." Tracker started, now entirely in control. "This area is much like my tribal lands. Lush but hostile, gullies and mountains." He pointed to the map to the northern part of the valley. "I could see that this was a non-defensible position and that the town could be taken quickly. As we went farther south into the valley, the mountain ridges ebbed and flowed from side to side, beginning just south of the town." He pointed on the map. "The valley began to narrow. The bridge over this river bed is like the stream bridge on our side, but smaller, maybe a hundred feet long, made of rock and wood. That's where we went into a hole."

"You're right!" Foxy called out. "I was looking off to the east at the railroad bridge."

"And I flew into the bushes!"

"We've talked about this Tracker. Get over it!"

"I have a long memory, my friend." He smiled.

Foxy continued. "That railroad bridge is very close to the highway bridge." He pointed on the map. "The terrain around here had pretty thick foliage, trees, and bushes."

247

Tracker kicked in. "The thing about this area was the similarities to a valley in my land. Like mine, this valley has gently sloping foothills right up to the mountain, then steep rock almost straight up. We stopped a few times, and I walked up into the hills. The higher up I went, the fewer bushes and obstacles I met, but it was rich with tree canopy."

"What about the railroad bridge?"

"It was just like the one we blew up at the Stream, Rusty. All wood. I think it's about the same length as the road bridge."

"Seems easy enough. We just have to get there and get out safely." Tully said, leaning over the map. "What about this side of the Peak, is it as difficult as the west side?"

Tracker shifted his position to get a little closer to the map and pointed. "See how much closer we are to the valley floor on this side. It looks worse than our climb, but it's not. I scouted this area and was surprised how easy it was if you knew how. Straight up from the valley floor is almost impossible. But if you hug the mountainside down into the foothills for about a mile, going north, the going gets easy. Another mile at this angle and you're on the valley floor. That will bring you out here," he pointed on the map, "about a mile south of the bridges."

Tully shook his head, pointed to the eastern edge of the Peak, "What about this drop from here? It looks like suicide!" It was a straight drop of maybe twenty feet onto a very narrow ledge.

"I thought so too until I repelled down to the ledge. You can't see it from here, but that ledge is not narrow, actually it's maybe two and a half feet wide, it goes deeper into the mountain then you can see from here. It's easy to walk on; then it slopes down into the foothills. But the first twenty feet require ropes, which we have."

Rusty smiled, "Good job, Tracker! And hey, lay off the mushrooms. Now let's come up with a plan for tonite."

Chapter Thirty Nine

"As I said, Jim, the bridges stand out the most, but we did discover something else nearby. About a mile south of them is a house and maybe a barn that showed three jeeps and at least three trucks in front. There appear to be guards around the perimeter."

"That's interesting, but what's so special about it?"

"Well, the picture isn't that sharp, but one of the jeeps appears to have a flag on the front right fender. My guys tell me that usually means a general."

"What side of the road? Any antennas?"

"Couldn't see any radio stuff, too much tree canopy. It's on the east side, off the road maybe it's in about two hundred yards. Rail tracks are back farther east beyond the compound. There's another structure behind the house, maybe a barn."

"What about the other valley, the Uijongbu Corridor?"

"Just started processing those films, should have them late tonight."

Colonel Dietrich continued to give him more details about the Cheorwon Valley.

Korea - The Peak –1730 – (5:30 pm)

Rusty wanted a single team. Tully wanted two, each taking a bridge. Red wanted three, the third acting as a rearguard so that the teams couldn't be cut off. Taking a break, they finally decided to delay any decision until receiving an update from Colonel Nelson.

"Roger, contact confirmed." Larry fiddled with the sending unit. Rusty nodded.

"Okay, ops for targets tonight. Leave 1900. Any update? R."

Hearing return clicks, Joe said, "Standby."

Click, click, click; He began translating. "Heavy road traffic. Approaches clear. Maybe HQ house one mile south of the bridge, east, between road and track. Peak Two is clear if need, but NK advance is now further south. No assistance available. Advise results. Good luck. N

Silence at the end of the transmission.

"Jesus Christ!" Red swore, "Don't we have enough on our plate?"

Rusty ignored Red. Leaned over to Larry, "Have you finished with the C-4 packs yet?"

"I've got two big bundles for each bridge all set to go, but I need to know how much time you want before I select the pencil detonators. I've got 15, 30, and 60-minute fuses."

Standing nearby, Arron asked, "What kind of fuse is that? Aren't we gonna use wire?"

"How long you been in the Army?" Red was still upset. They had to decide about timing, and now this new development. A fucking general's HQ!

Sensing the tension, Arron said, "I can hardly get my head around the latest in cooking technology, and you expect me to know about atomic stuff!"

"Sorry, Arron, didn't mean to put you down. I do admire your cooking technology knowledge." Rusty had shared with Red Arron's *sock* coffee cooking recipe.

Rusty was in a sour mood, as he felt put upon by the circumstances, but knew he had to take charge of the discussion. "Arron, the Brits designed and used these remote detonators in the War, very effectively. The attempt on Hitler's life in 1944 was a prime example of its use. A simple two-part chemical tube with a detonator. Break the vial and the chemicals mix, eating away the neutral wire. The chemicals are premixed based on the time needed to eat the wire. The whole thing is the size of a pencil, easy to use, and pretty time accurate."

Thinking of the open mission question, Rusty said, "Larry, I'll get back to you on the timing. What I want you to do is make up a few more packs, not real big but enough to blow up a house and railway track. I think we may need to create some diversions. Do you have enough C-4?"

"Got enough to blow up a few more bridges," Larry replied happily.

Rusty waved Red and Tully outside. "Let's finalize our plans. Arron! Get Foxy and Tracker."

Each had a map book on their lap when they began—so many unknowns, so many what-ifs. Everybody had a different idea or raised a concern. Tully finally had had enough, got up, and started pacing.

"You're the boss, Rusty, and I'll follow you into hell, so what you say goes. But we need to finalize this. I'd like to propose a plan. Please hear me out."

Rusty nodded, "You seem to come up with pretty good plans! Go ahead. What's your idea!"

"Our priority? The bridges. This HQ house is secondary, and maybe it's nothing at all. But it's the most dangerous. It's guarded, with how many men? A lot, I suspect." He glanced at Rusty for his reaction.

"Go on."

"So now I go to Red's idea of three teams. Road Bridge, Rail Bridge, and backup. But the Rail team also will investigate this general's HQ house, with the backup team on the west side of the

251

road to the house, just north of the HQ house entrance, as a potential assault force."

Rusty nodded, "So far, so good. Do you split the team for the house and the rail bridge?"

"That's the right question. I think the rail team crosses the road north of the HQ house, where the backup team will set up on the west side of the road. They'll cover the north and south traffic and watch the entrance road to the general's HQ."

"The Rail team proceeds to the tracks, but only one guy is sent south to the rear of the house to recon it. The rest of the team goes to the bridge, then comes back down the tracks to the house, finds out what's up, and either leaves quietly or blows the place up."

Red snapped. "That leaves a lot to timing, and good luck, Tully! I don't believe in good luck. So what's your thinking on timing?"

Rusty piled on. "That's what I'm struggling with."

Tully stopped pacing. "Damn sure those bridges are pretty well guarded, so don't think we have many choices. We're going to have to kill the guards at the bridges to get the charges set. How long before someone notices that they're missing? Not long, I suspect. So we have to set the timers pretty short, fifteen minutes at most. So even if the guards are discovered, they won't have time to search and find the charges."

Red said, "That's not enough time to get back to the HQ house without the detonations warning them."

"Right." Rusty had a thought and jumped in. "It's called a diversion! Any other way jeopardizes the bridges, and that's our main priority."

Red was unfazed. "This HQ house bothers me, Rusty. It's right across from our access up to the Peak. If we make a ruckus there, which, as things usually work out, we will, they'll be on us like flies on shit."

"Don't you think blowing up a few bridges might stir them up a mite?" Rusty fired back.

Arron was surprisingly quick on the uptake. "Crap! We haven't even spent a night on the Peak, and we're gonna move!"

Rusty was quick to respond. "Once you get past that, our decision is clear." He looked at Red than Tully. Each nodded agreement.

"That was a tough call, but you're right, Rusty."

"We're all right, Red! I'm just righter!"

He smiled to himself for making a small joke. "Let's pick the teams. Tully! Your thoughts?"

Tully resumed his pacing. Rusty watched him and pondered about how men think. Some like me, get quiet. Some like Red need to be loud. And then there were the walkers. Decisions are mind benders he thought, right or left, advance or retreat. A lot like playing craps, he thought, luck most times.

"Rusty, you lead team A for the road bridge. Your team would be, Frenchy, Larry, Snake, and Mario."

"Go on."

"I lead team B for the Rail Bridge and HQ house. I would like Foxy, Joe, Tracker, and Arron."

"Team C is Red, with Dean and Moose to protect our asses."

Rusty wasn't sure. "So you're thinking of leaving the Peak guarded with Marty and Ben? I disagree, I don't think it needs to be guarded at all, and we could use them. What do you say, Red?"

"Have to agree with you. Chances of the NKs getting to the Peak before we get back are slim."

"Red, what's your thought about the team makeup."

"You know me, Rusty, I tell it as I see it. It's a good plan, but I have some suggestions."

Tully interjected. "Red, you saved my ass on that last OP, so let it fly. I have no ego here."

"I'm not the smartest guy, but I've seen more than my share of foul-ups and disasters, so I tend to be on the cautious side."

"You saved a lot of our asses, Red. Just say what's on your mind." Rusty said.

"The road bridge has a lot of traffic, so the chances of taking out the guards unnoticed seems pretty high. I suggest we add sniper Dean to Rusty's team to protect the demolition team and the explosives. He's our insurance policy."

"Next, I would add two more men to this NK's general HQ house recon. Too many guards there, the explosions will alert the house compound before the full team returns. This three-man team would be tasked to take action immediately following the bridge explosions."

"Lastly, I want to prepare surprises for anyone trying to follow us up the trail to the Peak, because I think they will."

Tully smiled. "I like it, Red. Just goes to show you two and a half minds are better than one, with me being the half."

Red was amazed. "Tully! I never heard a Ranger take a step down for anybody. Are you becoming human-like?"

"Sometimes we Rangers reappraise. It makes us more Godlike, don't you think?" Tully grinned.

They decided on the final teams, timing, and details of each. As the men broke to prepare for the mission, Tully pulled Rusty aside.

"Where are we going next? Is it Peak Two? How far?"

"Yes. Ten miles south, near our depot # 2."

"Tell me exactly where it is, you know, in case something should happen."

Rusty was a realist. The question didn't bother him and had expected it. He pulled out his map, showed Tully approximately where the depot was, and about where Peak Two would be.

"Only Red, Arron, Foxy, Tracker, and I know how to get up to this hideout. It's well hidden. Even if I told you exactly how to find it, I doubt that you could. So, make sure one of us survives to show you."

Chapter Forty

Everything had been prepared for a hasty retreat if the mission went sour. The sun was setting and cast a long shadow on the eastern slope they would now descend. Red and Arron lowered the heavy gear by rope down to the ledge, below the peak. The men followed, repelling down, one by one.

Tracker led the way down the ledge, Snake, then Rusty. Arron followed carrying half of the bazooka and bitching about its weight, while Moose behind him, took his heavy BAR and the other half of the bazooka, not saying a word. Red was last in line and carried the machine gun. All the men took extra everything with stuffed backpacks; extra ammo, grenades, C-4 charges, bazooka rockets, and machine-gun belts.

It was rough going, but they held a steady pace. Rockslides over the years left the pathway deeply rutted in places, with drops of two feet or more. Fauna and odd branches grew out of the mountainside, making the men duck or carefully navigate around the obstacles.

Not too far down, Larry stopped, took something out of his jacket.

"What's going on?" Red asked from behind him.

"You wanted surprises, right? I'm picking spots now, so I don't have to feel my way in the dark." With that said, he tied a white piece of cloth at the tip of a small branch growing out of the path wall. Red watched and knew what Larry was planning.

"Glad someone listens to me!" Red said.

Along the way, Larry set seven more cloth markers.

It took an hour for the men to reach the beginning of the foothills. The sun was just going down over the horizon and cast the lingering glow off of the clouds that made the surroundings foreboding, like a haunted forest, there was enough grey light to navigate through the underbrush. Their pace quickened.

Rusty worried about the soon to be rising crescent moon. Would it be too bright? What about rain? Would it come later? If it did, could they get back up that path in pitch dark with heavy water runoff from the mountain? Okay, he said to himself, I'm praying to you, my unknown spiritual protector. Please keep us all safe tonight!

Suddenly, Tracker stopped. "Snake! Stay here with the column; I'm going to scout to the road."

The men could hear the muffled engine noise from the trucks passing two hundred yards away. The undergrowth to the road was much thicker and made for ideal cover. The moon was now just peeking out just above the eastern mountains casting its small amber light unimpeded by clouds.

Rusty moved closer as Tracker returned. "We're just south of that general's HQ house entrance. There's a good spot for Red to set up."

"Are you sure? Could you see the house?"

"No, it must be too far back. But I did see an entrance path off the highway, with a jeep to the side and two guards. Nothing else it could be."

"Road traffic?"

"Mostly trucks, moving steadily, spaced pretty far apart. Shouldn't be hard to slip across."

"Good job. Get Red's team set up before you cross the road."

Red and Tully knelt next to Rusty for the final briefing.

"Remember, bridges first. Okay, it's 2100 now," they synchronized their watches, Rusty continued, "We blow the bridges at 2300 (11:00 pm). Anybody disagree?"

"Two hours should give us plenty of time," Tully said, Red agreed.

"I'm heading out. Good luck!" Rusty got his team going with Snake doing recon out front.

As Red set up his position, Tracker rested off to the side with Tully and his team.

Tully's head banged with the word of concern that Red and Rusty voiced. Timing. He thought Red was right. This damn HQ house was the weakest aspect of the whole mission. Was it worth the risk of killing a general? His mind sought justification. What if we could've killed Rommel in North Africa before he fortified Normandy? I saw firsthand what this one general did. He probably cost us ten thousand lives. Who the hell is this guy in this house? Is he that important?

He refocused. You know your mission, Ranger! There are no certainties, are there? Just go with your instincts. You've done this. Do it again!

Tully considered the change in the teams. He liked the new setup. It was better this way. His team was critical, though. It always seemed to work out that way, didn't it? Yeah, he must be the luckiest, right? But now, he was the one responsible for the success of the mission. Don't whitewash this, Tully. You're accountable for the lives of fifteen men here. Yes, it's on me, but damn! I have a team to do it!

Chapter Forty One

Monday - June 26, 1950,   Day Two of the Invasion
Korea – Somewhere South of Kaesong on the Western Invasion
Route – 2100 - (9:00 pm)

George had gained his bearings that morning, and realized he was just a few miles north of the Han River intersection, on the bank of the Bukhan River. If he could cross the river, it opened the door to Seoul and his survival.

After flowing around the south side of Seoul, the Han River turned north for almost thirty miles, where it than met the Bukhan River and then proceeded out to sea.

From his hiding place, he witnessed a masterful invasion by the enemy. A well-coordinated artillery barrage followed by rubber boat crossings to secure a beachhead on the far bank. Then the big trucks came with their river bridge crossing equipment and pontoon boats. Under constant shelling, they built the bridge across the river. These guys are impressive, he thought. They have good stuff, and they fight. Shit! Kind of like the Japs but better equipped and no fanatical crazy charges. They performed like a well-oiled machine.

By midafternoon, the enemy started crossing. George was thinking his only hope of crossing safely was by that bridge. Watching all day, and now early night, he realized that the enemy was feeling safe crossing because there was no more shelling.

 The fighting had rapidly moved further south. As darkness descended, his escape route was cast in the ambient light of the vehicles crossing this pontoon bridge. Troops walked the sides, next to the vehicles.

Okay, now is the time, he thought. It's dark, lots of noise, no one is looking in the river. Let's go, George! He crawled to the river bank, slipped in, made his way as far out as he could and relaxed, spread flat, and let the river take him toward the pontoon bridge. He floated, not wanting to make any noise or splashes. The current was strong and carried him swiftly. The lights from the vehicles glowed on the river, showing him as a target.

The line of bridge support pontoons that stretched across the river were spaced very close together, needing to support the heavy T-34 tanks. They were big hard rubber inflatable things that supported prefabricated aluminum bridge sections that just bolted together.

Rapidly floating towards this water crossing structure, George's survival depended on being able to grab hold of something, and then....well, what to do next? Something always shows up, he thought. It always did. And it did this time, too.

He hit the big rubber pontoon with a thud, started grabbing at the sides, and got a grip on something. The two-knot current turned him sideways, but he was able to hold on. He was between two of the floating supports with maybe a foot between them. The rumbling four feet above him was disturbing, and the cold river water was taking its effects; he started to shiver.

The something he was holding onto turned out to be a rubber cleat that he somehow managed to use to lift himself enough to grab a support beam that held the bridge section to the pontoon. Swinging onto the top of the rubber side, he laid flat and took a breath. That was fun, he thought, breathing hard, feeling how close he was to being dead.

Now the going got easier. Grabbing the overhead I beams and swinging between the pontoons, with about two feet of clearance to the roadbed above, George had plenty of room to move fast.

He got near the far bank in time to see thick clouds moving in. Thank God, he thought. Slipping back into the water near the end, he followed the shoreline away from all the traffic coming off the bridge until he felt safe.

It was pitch blackness now, with flashes off in the distance. He guessed a few miles away.

A few hours later, George had made his way southwest to the edge of the northern section of the Han River, south of where it intersected the river he had just crossed.

I'll follow this sandy bank all the way to Seoul, he thought. But then he had second thoughts. It's clearly the easiest route for me, but wouldn't the enemy think it as well?

But he thought, so far the NK's are going straight down the coast highway. Why would they come a mile out of their way to get over here? George was concerned about the gap between the coast highway and the river. He correctly assumed that nobody but a desperate attack force would try to use this incredibly tough terrain to move south.

Well, they might if the ROK held them up somewhere farther down the road. Okay, I'm good with that, I'll keep an eye out for a prolonged engagement.

While narrow, the river bed was ideal for a small group or, in his case, a one-man band, to move quickly and safely.

Twenty miles and counting, he said in his head and continued his conversation. I never thought I'd say this, but damn I miss the jungle. This country is shit! He was angry, he had to be so exposed and so near stumbling into God knows what. You're alive, buddy! Stop complaining. You're making progress. I figure I could go three miles in two hours. Maybe I can do it in one hour.

Moving rapidly at first, he then encountered heavy shelling and a fierce engagement nearby. He stopped to assess whether to go any further. His instincts were well developed and proved to be a lifesaver again.

Within minutes, soldiers came down the bank just in front of him. He could only hear them, as they made a lot of noise. These were regular infantry with full gear. He didn't know how many but guessed it was a platoon, maybe thirty men as they moved south along the bank. He didn't move.

Sticking to his plan, he waited ten minutes then decided to follow. George slowly moved another five minutes or so and went flat when heavy machine gunfire erupted in the near distance. It sounded like a Browning fifty caliber machine gun. God bless America! He said in his head. A short exchange took place with some grenade or mortar explosions intermingled with small arms fire. Then nothing.

Shit! He wondered who won. Okay, now I know I'll get killed here by either side if I mess up. He thought of his brother just then. He always said, before you jump into a bar fight, wait till you know who's gonna win. Okay, I'll wait a bit, but I think I already know who's winning. I just gotta get the hell out of here.

Chapter Forty Two

Monday - June 26, 1950, Day Two of the Invasion
Korea - The Bridge Mission
The Road Cover Position - 2110 – (9:10 pm)

"This is a great position," whispered Ben.

It was elevated, maybe ten yards off the road with a firing lane of a hundred and eighty degrees, and had excellent cover. Low rock formation allowed the machine gun to could be fired from a prone position.

"Damn, this is good luck! A good start." Said Red.

Marty had carried the tripod and finished setting it in place; Red set the weapon. The three of them removed the ammo belts from around their necks and emptied their backpacks of more belts. They were set. Red watched trucks slowly pass. Their low beams were on, and they had large gaps between them.

Yeah, Red thought, this couldn't be more perfect. That general's HQ house entrance road is visible across from them. As a truck passed, it cast a light on the jeep stationed there but could only see one guard just off the main road. He couldn't see the other one Tracker mentioned.

Marty loaded the first belt. He didn't have to be too quiet; the noise of the passing truck hid the sound. He spoke up so Ben and Red could hear him but still in a hushed voice, "Just want to remind you guys that I'm not just a combat medic. I've killed many, I've survived hand to hand combat, and I've never let my fellow soldiers down."

Red looked strangely at him. "What the hell! Is this a Jewish thing?"

Marty got flustered and couldn't respond.

"God damn it, Marty!" Red angularly whispered. "Don't you know you're the most respected member of the Post. That's why you get the safest duties. You earned your combat status. We want you to be safe so you can save our asses. So cut your shit. Besides, I think you have more medals than I do."

Having heard the whole discussion, Ben leaned over Marty to get closer, "You broke my curse, Marty! Patching up Manny and saving him. I'll never forget that. You're part of my family now."

Red signaled Tully that he was ready.

The Road Crossing – 2120 – (9:20 pm)

Tully acknowledged Red's okay and led the team closer, behind bushes ten feet off the road. Tully had been timing the passing trucks and looking at the HQ entrance about two hundred feet away, making sure they wouldn't be highlighted by the vehicle headlights when they crossed or by the cast in the weak moonlight. No, he concluded, not a chance of that happening. Even the universe cooperated by a passing cloud, dimming the small moon's increasingly brighter glow.

As prearranged, Tracker and Foxy would be first to cross. Tully ordered. "Go!." Joe and Arron were next. They watched as the two men scampered across and dove into bushes on the other side. They waited. No alarm. Nobody saw them. The next truck passed, and they crossed.

Tully was feeling better. So far, so good. He was next, with Moose and Mario. He leaned over to Moose and whispered, "You want me to take that bazooka half you're carrying?"

Moose turned his head to face Tully and gave him an odd look and said, "Might slow you down, I'm good."

The Road Bridge – 2150 – (9:50 pm)

Rusty's team made slow but steady progress. Snake found a ditch that paralleled the road to the bridge. It was close enough to hear the traffic but still far enough so that anybody on the road couldn't hear them. The mile-long journey to the bridge took forty-five minutes. Rusty checked his watch. Good, he thought. Fifty minutes plus to set the charges. Plenty of time to figure out how to do this.

"Snake, recon the area. See if you can get close to the bridge. Check things out for Larry."

Snake nodded, dropped his Tommy gun, gear, and helmet, and moved out.

As planned, Dean started searching for his insurance sniper spot, and within a few minutes, found the tree. He wanted to be able to stand and have a proper angle with his rifle being held steady. The V in the branches of this tree was ideal, shoulder high, unobstructed view of the entire bridge. He set up his scope and peered through the lens, made some adjustments, and settled in.

As they waited for Snake to return, Rusty passed his binoculars around to each of the team to survey the area, what they saw was daunting.

A guard was stationed on each side of Rusty's end of the bridge exit, twenty-five feet apart. Another set of guards were on the far side, one hundred feet away. Most disturbing was the emplaced sandbag position down on the west side of the river bed, ten feet from the arch access under the bridge. They couldn't see the other side but assumed there was a similar emplacement.

Rusty mulled this new development. Not so stupid, these Commies. I guess they understand the importance of these bridges. His mind screamed. Have you ever met an ignorant enemy! Okay! I know! Plan for the worst and hope for the best. Think! And think fast, buddy. You've got less than forty-five minutes to execute this plan.

Snake was soaked when he returned like he'd gone for a swim. Dean was watching, guarding. The rest of the men came around as he sat with Rusty.

"I found another enemy position downstream of the bridge, about two hundred yards west, just off of the river bank. After seeing what the positions were at the bridge, I thought maybe we could use an old technique smugglers used to cross the Rio Grande near my town safely. They got US Army uniforms, the same as that of the troops that patrolled the border. They rarely got caught. If they did, it was because they couldn't speak English. Anyway, I have two dead naked guys back there, and two slightly stained uniforms here that I think solve a lot of problems."

"You continually amaze me, Snake." Rusty realized that Snake had come up with the solution. "Did you see any patrols?"

"No. But I only scouted the west side."

"You really solved our problem here, Snake. I'm gonna put you in for a medal. Of course, that's if we get out of here."

"I'd rather you just get us out of here."

Rusty rechecked his watch, "Okay, everybody, it's 2215 (10:15), check your watches; we've got thirty minutes to set the timers. We must move now."

"Snake, you and Frenchy get these uniforms on. Larry, you're going to become a prisoner of the two cross-dressers when they show up." Rusty paused and looked around at his men and said with determination, "The curtain is going up, guys, so play your parts and stay calm. We'll do this, and then we'll get the hell out of here."

Rusty went back to Dean and filled him in on the plan.

The Split-Up by the Tracks – 2130 – (9:30 pm)

Tracker led Tully's team toward the railroad tracks along the outside of the general's house compound perimeter. Past the house, an outline of another small structure appeared under trees away from the main house. It looked like a barn. They assembled in a small clearing nearby. The moon shed some light through clouds, enough to navigate a path but not enough to define precise shapes.

Tully knelt, "Time to split-up here, this barn or whatever it is, is unknown. Foxy, you're in charge of the HQ's team, deal with it. We talked about this, but I just want to remind you of how important the bridges are and that this HQ house is secondary."

"Back off, Tully! Foxy knows the plan." Joe whispered tensely behind Tully. He was aware that Foxy was totally aware of all its implications.

"I'm good, Tully. Joe says it straight." Foxy turned to Joe. "Thanks for your support, but it's not needed."

"You're wrong, Foxy; help is always needed."

The Railroad Bridge – 2200 – (10:00 pm)

The tracks turned east at an angle after crossing the bridge and then veered west heading south, following the road south pretty closely.

With Joe on point, the railroad bridge team moved up the tracks, Tully followed, with Arron and Moose trailing. It was less than a mile away from the general's HQ house. With no patrols, they made it to the crest of a small rise overlooking the bridge in thirty minutes.

About two hundred feet from the bridge, they found a place that had some cover. Two guards, both smoking, were visible, sitting off to one side on a boulder. Joe went off to scout the best approach to take the guards.

Tully nodded to Arron to set up the bazooka. They had a clear shot at the road bridge from this spot. Red had insisted that they take the weapon, another insurance policy in case something went wrong at the road bridge, and that team was discovered before they could set the charges.

Moose and Arron put the bazooka tube together, set the targeting device, loaded a rocket, and armed it.

Fifteen minutes later, Joe was back. "No patrols, only these two guards. There's a ravine to the east of here that runs right behind these guys to the river bank. Should be easy taking them out." Tully checked his watch.

"We've got forty-five minutes to set the timers; we leave in fifteen." Joe gave him a thumbs up.

The general's HQ House – 2130 – (9:30 pm)

Foxy's team cautiously approached the HQ property. As planned, Mario would recon the north side of the house perimeter, starting at the outside edge, and follow it to the road. Then he'd move back, trying to get closer to the house if possible, then check the rear, and finally return to the tracks where he started, and report.

Tracker would follow Mario's same plan but on the south side of the property. Foxy would investigate the barn.

Foxy scouted the north side edge of the barn almost to the front, then came to within ten feet of the side, stopped and scanned through the underbrush. No guards, no door, and no window. Man, it's old, he thought, and in deep need of repair, small too, for a barn. But I'm no Korean farmer. Crap, I'm no farmer period, what do I know. He did see a faint yellow glow of light coming through wall plank seams that had separated with age. Ah, maybe someone's inside, better check around. He went back into the bushes and circled to the rear of the structure.

At the rear, no guards, no door, but a window. Well, not really a window, but an opening with a closed shutter. Maybe two feet square, he figured, five feet off the ground, probably for ventilation. The shutter was closed but was missing a board; the inside light shone through. Okay, when I finish checking the other side, I'm gonna have a look.

He turned to withdraw back into the brush when he heard a piercing scream from inside the barn. He froze. Only once in his life had he ever heard a scream like that. No! Please no! He started to

267

sweat. The agony in the screams continued, he heard muffled yelling. The screams continued.

He prayed. Holy Mother of God, give me the strength to bear this. Please let whoever is being tortured die quickly. I can't help now, but I promise I will end it, and I will not be merciful, like you, my Lord.

He forced himself to move and scout the south side. Away from that rear window opening where the sounds were now muted but continuous. Nothing different on this side, the same as the north. He kept focused, trying to block out his thoughts of what was probably going on inside that barn.

No longer hearing any sounds, he moved to the front of the barn. About fifty feet away stood the main house with two guards at the rear porch door. Oil lamps were on in the house. Two guards had just passed, apparently walking the house perimeter. He did a quick look around the corner of the barn and saw two guards standing at a big door. He'd seen enough and headed to the rear. Can't stop now, he was late and needed to get back, all the while he was thinking about what kind of place this was, who is in that big house and how was he going to kill those bastards in the barn.

At the Tracks behind the Barn – 2145 – (9:45 pm)

Near their meeting place back by the tracks, Foxy was last to arrive.

Tracker caught Foxy's face. "What's wrong?"

"There's some horror going on in the barn, Tracker. You know, like Nazi SS shit. What's the house look like?"

Mario was no stranger to what the SS did. "Did you see who's in there?"

"No, I wanted your recon first before I attempted to take a chance and see what's going on. Don't have too much time left before we can do anything."

Foxy checked his watch. "One hour and fifteen minutes to detonation. What about the house?"

Mario's eyes widened as he spoke.

"There are two guards just off the main road, with a jeep nearby. The entrance road skirts the northern edge of the property to about one hundred feet from the house, then turns sharply south into a mostly open field in front of the house. I saw three big trucks randomly parked and three jeeps in this area, plus a lot of troops, maybe a platoon, scattered around. These guys are not expecting anything. I saw the guards laughing and smoking, enjoying themselves. The house is well guarded, four at the front and a continuous two-guard walk around. Two guards in the back of the house and two guards at the front of the barn. No guards along the perimeter. I couldn't get close enough to see inside."

Foxy looked at Tracker. "No perimeter guards. I agree with Mario; I think a platoon. Twelve guards showing and I counted fourteen soldiers, some sleeping, some milling around the trucks, mostly settling in small groups. Must be a few in the house and barn, plus officers. Can only guess how many officers?"

"We need to know about the barn. Tracker, check out the rear opening, Mario, you've got his back. Remember, no action. I've got to plan this. We need to act. Be back quickly."

They left Foxy to his thoughts. Red was right; this is going to be ugly. It's ugly now! Think! He tried to calm himself. I don't give a shit about any enemy general. But orders? Right! I know. Okay, the barn. My mission, no, our mission. Foxy! Stop messing with yourself. Accept it. You can't save everybody no matter how hard you try, can you? Just get on with it!

And he did. His plan became clearer but with many doubts.

Behind the Barn – 2150 – (9:50 pm)

Under the window opening behind the barn, Tracker heard someone talking. In English! "What the hell," He said under his breath. The screams inside suddenly started.

269

He flinched, then straightened up to see through the opening in the shutter. "Jesus Christ!" he almost said out loud. He ducked instinctively, thinking someone inside might have heard him. No one did. He went back to look.

His elders had told him on the reservation of the techniques their warriors had used to extract information from the White trespassers, mostly though, they were used for revenge and also to warn other White folks off. He had been outraged by the tales of savagery told. He was witnessing this same cruelty now. Yes, he thought, I've been savage. But never on tied up soldiers. I've never butchered anyone. His blood boiled at what he was seeing.

His view into the barn was unobstructed through the slit. Near the front of the barn, on the south side of the front door, was a table with an oil lamp and three chairs where two officers sat. Ten feet away was a center post with a soldier tied to it, facing the officer's at the table, Tracker couldn't see the man's face. He must be special, he thought, as his interrogators had added a crossbeam to hold his arms out, as in a crucifixion. "You animals," he whispered.

He saw the reason for the screams. The outstretched left hand was bleeding badly. It was missing three fingers. The officer in front of him was talking to him, in English, while holding a wire cutter.

"I do enjoy doing this, sergeant! I suppose you don't. I will stop, I promise. If only you'll tell me about your unit and its mission. You're not an advisor or trainer, so who are you? Why were you in Cheorwon?"

Tracker thought he heard, "Go fuck yourself."

"You are stupid. You let one man die already. Do you want this poor bastard on your conscience as well?

Tracker heard the response from a voice filled with pain and defiance. "You're gonna kill us all anyway; we're just trainers, that's all."

"No! You're not!" He pulled out his revolver and walked over to the north side wall where a soldier was lying. Tracker hadn't seen him before.

"One more chance before your friend dies!" He paused, "No!" He raised his pistol and looked back at the man, "You sure?" He paused for effect. "Okay!" He fired.

The man on the cross screamed almost incoherently, "You are going to burn in hell!"

Tracker saw a third man hanging from a rafter in the far corner. He was skinned. His legs, his arms, were raw. He was dead. Tracker shivered.

An officer at the table spoke English, in a heavy Russian accent, "We've learned nothing from these Americans. Who cares anyway? What could they possibly do now? By the time they wake up, Korea will be united."

Tracker had heard enough. Angry as he had never been before, he grabbed Mario and left.

"God Almighty! I almost had a heart attack when I heard that shot. What happened?" Foxy's eyes were wide.

The three huddled.

"Evil business! Like you said. American soldiers in there. I think there's only one alive. But he is hurt very badly. They killed one as I watched, right in front of me. That's the shot you heard. Two dead US soldiers, terrible. Three officers in there. Two North Korean, one Russian. I didn't see a general. No guards inside."

"No guard came in after the shot?"

"No."

Foxy nodded. "We've got less than thirty minutes to the big noise. Here's what I want to do."

271

Chapter Forty Three

Monday - June 26, 1950   Day Two of the Invasion
Korea - The Bridge Mission
The Road Bridge -2230 – (10:30 pm)

Frenchy and Snake were not big men. But they looked ridiculous in the much smaller North Korean uniforms. In the dark, helmets on and with arms raised holding their rifles to hide the shortness of the sleeves; they figured they'd pass a cursory look.

"You okay, Frenchy?"

"Oui, mon Général! I will play a wonderful part. The enemy, she is ours."

"Speak Korean, you idiot, or we all die," Snake said at his side. This was something new to him. Exposed. He was used to being the ghost! He sneaked up on the enemy, didn't like this new role one bit.

Larry followed them. When they dropped close to the riverbank, he moved to their front. They approached the gun position near the bridge with Larry's arms raised; their rifles pointed at his back. As they drew near, Frenchy yelled in Korean. "Prisoner here, don't shoot!"

One of the guards called back. "Who are you? What unit? Stop where you are!"

They didn't stop, Frenchy responded, "We are a special security unit of General Choi Mo. We have captured an American." He made up the name of a general, hoped nobody knew any.

The guards at the river edge of the bridge were relaxed, standing with rifles at port arms. Frenchy's ad-lib about the general was enough of a diversion to get them close enough to allow Snake, holding his rifle straight in his left hand, throw his knives with his right hand, and did with deadly and silent effect. Thump, thump. Two dead guards slumped to the side.

Frenchy and Snake took over the guard's position.

As trucks rumbled by, the guard above at the end of the bridge heard nothing. He wasn't looking anyway. He'd been on duty for the last seven hours.

When they had settled in, Frenchy leaned over, whispered to Snake. "You're good! Nice job."

Snake leaned back and grinned, "Merci!"

Larry unslung one of his backpacks and quickly moved to the bridge at the river's edge. Thank God, he said to himself, as he saw boulders above the rushing river under the arch. The moonlight was dim but just enough. He moved out onto them. The noise from the rushing river was amplified under the bridge, intensely loud. He walked along the edge to the center of the buttress.

He had to make sure there wasn't another guard emplacement on the eastern side of the bridge as well. Didn't think there would be, and there wasn't. Relieved, he moved back along the edge to the center and felt around along the wall. Finding an outcropping, he hung one of his loose packs on it. Damn! He thought, wish I could put my second charge out in the middle of the arch over the river. Probably don't need it but it would be proper insurance. I'll feel around? I wish I could use my flashlight.

Feeling his way around the underside of the bridge, he found a steel reinforcing beam that must have been added by the Japs, ten or twenty years ago. Shit! If I had missed this, this bridge might not have come down. Good job, Larry! He then shimmied out to the center under the arch on this beam. Hanging with one hand, he undid his other backpack and strapped the charge to the girder. He checked his watch.

273

"Shit! You've got to be kidding!" He said, knowing he'd have to hang there for three more minutes before he could set the timers.

Outside, Frenchy and Snake had assumed the position of the two dead guards and tried to act casually, as they might have, kind of huddled, smoking. Frenchy was making small talk in Korean, with two different low voices. They worried the guard above at the bridge might get suspicious. He was close enough to hear them.

The road traffic was getting lighter, Frenchy noticed. Not good, he thought.

The Railroad Bridge -2230 – (10:30 pm)

Joe and Tully entered the gully not far off the tracks and moved forward. The guards had now separated, standing at each corner of the bridge, across from one another. Joe took the lead and came up behind the eastern guard. With a swift grab and snap, the guard fell like a sack. The western guard heard the noise and turned, moving his rifle off his shoulder. Tully had come out with his Bowie knife, and with a quick flick of his wrist, the knife went deep into the guard's chest.

They hid the bodies to the side, then moved underneath the bridge. Tully climbed onto to the crossbeams and moved out towards the middle, walking the beams and holding on just below the tracks above, swinging one arm, then a leg, making good progress.

Joe stayed at the base and located a place where the bridge was anchored to the river bank with concrete and steel. He removed his backpack and set the C-4 package, checked his watch.

Tully got to the center, found a good grip where he could hold on with one hand and removed his pack. It was easy to place the charge and secure it; then he checked his watch.

At the same moment, Arron was scanning the road bridge with his binoculars as he had for the last thirty minutes and noticed that the traffic had gotten lighter. Shifting his focus to the north end of

274

the road bridge, he no longer saw headlights. "What the...?" Then he too checked his watch.

The Road Bridge – 2243 – (10:43 pm)

Dean watched the bridge guard above Frenchy's position, wondering if he was going to have to shoot this jerk. He thought the guys could handle it without him alerting the world, so he just scanned the area as he usually would. The last truck in line was just finishing the bridge crossing, so he focused on the far side. "Holy shit!" He let out in a low voice, for what he spied through his scope on the far end of the bridge shocked him.

In rows, four abreast, troops started crossing the bridge. They looked relaxed, walking leisurely in loose formation. It was a long column, and Dean couldn't see its end.

At the same time, Rusty saw them through his binoculars. He dropped his field glasses and raised his wrist, did a quick time calculation, and said, "*SHIT!*" He wanted to panic but grabbed hold of himself, talked himself down. You got trained people all around you. You planned well for almost anything, and it will soon take on a life of its own. Trust your men's instincts! Mostly trust your own. Go man! Do your thing!

And he did. He found Dean, told him what he wanted, and returned to his firing position.

Under the bridge, Larry's arms were killing him after hanging for two minutes. "Screw this," he said, thinking that in another minute he was gonna be in the water. Barely able to hold on with one hand, he broke both timer fuses in the C-4 charge. He scurried, hand over hand back to the second charge on the wall, and set those timers.

Moving quickly along the rocks, he came out from under the bridge, saw his guys sitting in the enemy position, and stopped. He heard no truck noise. The only sound was the rushing water echo from under the bridge. Something's wrong, he thought. He froze.

275

Snake saw him appear at the edge and slowly raised his hand to signal him to stay put.

The guard above had decided to take some interest in them and had asked something.

Frenchy responded playfully. Snake didn't like the feel of this and got ready. The guard seemed confused and said something else and, at the same time, turned from his post to look straight down on them.

Frenchy said something else, but this time the guard reacted by unslinging his rifle. Snake responded with one of his knives, and the guard fell backward onto the road in clear view of the east side guard.

This guard immediately brought his rifle to bear and rushed to his comrade. Bang!

Dean's shot rang out like a small town Church bell at eight in the morning.

The guard dropped, and then all hell broke loose.

Larry ran from the bridge with Snake and Frenchy directly towards Rusty's position. No evasion, they just ran like hell.

The troops on the bridge, now almost halfway across, dropped to a crouch and waddled to the protection of the stone bridge sides.

A few soldiers raised their heads and clearly saw the three men running on the far side. One drew his rifle up to shoot. Bang! His helmet flew off and clanked on the rock road surface, once, twice, and then rolled around, his body falling with a thud before the first clank. Nobody on the bridge moved.

Rusty watched all of this unfolding, prayed his bridge team would all make it back, and checked his watch again, 10:53, seven minutes. What now! No choice.

Frenchy was first to change uniforms and came close to Rusty. He was within earshot of Snake, who was also changing. Frenchy spoke, in a French accent, "I want to report considerable progress with your men's language skills."

Rusty was kind of preoccupied and looked strangely at Frenchy.

Snake had a small laugh.

On the far bank, Rusty watched as troops dispersed. Light mortar teams were starting to set up.

The troops on the bridge slowly advanced alongside its protective walls.

Rusty looked back at Dean. "Shoot a few on the bridge, and then we're out of here." He turned back to his team, "Snake, take the point, go!"

He saw an explosion on the far bank.

The Railroad Bridge – 2258 – (10:58 pm)

Arron was freaking out. "Oh, my God!" He said out loud to no one, "Must be a whole freaking regiment!" He rechecked his watch. No way is it going to blow before they cross. Why me? Best cook in the Army! Right!

He didn't want to remember the special operation in Greece in 1944, where he spent four months in the mountains evading the Nazis. The memory came to him, as it so often did. Life and death decisions come fast. That one young Nazi soldier he caught on patrol. I tried to save him until he turned on me.

That didn't work out too good for you, asshole, did it? My motto, he said to himself, as he did many, many times, just to keep his sanity, was, my food keeps us alive, my heart is pure and sustains me, and my gun protects me. Then he thought, thank you, God, for making these assholes northerners!

Completing their demolition placement on the railroad bridge, Tully and Joe rushed to Arron's side. Moose was positioned a little closer to the road bridge. Arron was ready with his bazooka.

A shot rang out. Then a second.

Tully checked his watch. "Jesus!"

Tully watched the enemy's reactions as troops moved out along the far bank and started shooting at Rusty's position. Mortar teams started setting up their weapons. He tapped Arron's shoulder, "Fire on that far bank against those mortars."

Arron moved his aim, "I'm not Tom Mix, you know."

"Just fire!" He did. *Swish*! The rocket struck a tree behind the enemy on the bank. Nice explosion. Just what Tully expected. Everybody on the far bank dived for cover.

A whistle blew, and the men on the bridge got up, started running to finish crossing to the south side.

Moose opened fire as soon as the troops started moving. His BAR jumped, soldiers began falling. The twenty round box quickly emptied, he jammed in another.

Tully had already loaded Arron's bazooka with the second to last rocket they brought and tapped Arron on his arm.

"Get the bridge, Arron."

Swish. The rocket went off into….nowhere. Tully quickly loaded the last rocket. Arron lifted the bazooka and aimed. Tully tapped his helmet.

Swish. Explosion. On the road, in front of the road bridge.

"Crap!" Arron let out.

It started getting intense. Troops on the far shores of both bridges had dispersed, and now the enemy began focusing on their position as well, with heavy fire. Moose's BAR kept mowing down troops trying to exit the road bridge, and after the first few rows of men were shredded, the enemy stayed down.

Joe could see soldiers scrambling at the far end of the rail bridge, crawling up onto the bridge tracks. Even though they were far away, he started shooting at them with his Tommy.

Tully watched mortar rounds impacting Rusty's position by the road bridge, and hoped the hell you're out of there, he thought.

"Moose! We gotta move now! He yelled. "Let's go!"

Arron turned and bent low, trying to hold on to the bazooka and run.

Moose came running up next to him and grabbed the bazooka out of his hands with one hand while holding his BAR in the other.

"Come on, Arron! No time to hunt for mushrooms!"

Arron relieved, he could run faster, "That wasn't me!"

Joe brought up the rear.

"KABOOM"! The explosion at the road bridge was loud. They were a hundred yards down the tracks and couldn't see it, but knew it must've done the job. Then a series of deafening explosions followed almost in sequence. Tully knew that his, and Joe's demolition bags had worked perfectly, as he ran down the tracks with a big smile, talking out loud to himself, "Larry! I owe you big! Maybe that's two times! But damn! You're fantastic!"

Chapter Forty Four

Monday - June 26, 1950,   Day Two of the Invasion
Korea - The House Mission
One Hour Earlier -The HQ House – 2200 – (10:00 pm)

Tracker and Mario started executing Foxy's plan. Teamed up, they crept along the south side perimeter past the barn and the house and got to the front area where the bulk of the troops were located. It was late, and many of the men were sleeping. They first crawled to the trucks and planted C-4 blocks under each, using Larry's sixty-minute pencil timers. They ignored the jeeps, knowing they couldn't get close enough. They weren't important anyway.

Next, they moved to the far southwest corner of the property, turned around, headed north, silently killing the sleeping soldiers or the not so sleeping soldiers. They established a grid pattern in their minds and were going to clear it, one grid at a time.

Their hearts were cold. Both had heard the screams from the barn. Both had seen what the Nazi SS had done to their fellow soldiers and innocent civilians. These animals were the same, Mario thought, show no mercy, get no mercy, ran through this mind.

Tracker was the most incensed and was deeply offended by what he witnessed. These people deserve to die and die badly. I wish to help them do that, he thought. He slashed and cut with a smile and a curse to this repulsive enemy.

The HQ House -2250 – (10:50 pm)

Tracker and Mario agreed to stop the grid pattern killing at a particular time and move to take out the front entrance guards by the road. Under a tight schedule, speed dictated the removal of the guards so they could move into position for the next act. Rechecking their watches, it was six minutes to showtime.

Foxy stayed under the barn's rear window, waiting. He was grateful that the officers had decided that drinking was better than doing anything more than throwing insults at their pathetic captive.

He got the impression that the Russian didn't speak Korean well, but had a good command of English. He also thought the other Korean officer didn't speak English at all but spoke Russian. The guy who liked to torture, yeah, he was very educated, spoke many languages. You're mine!

Foxy moved around to the front edge of the barn and took a quick look. The moonlight filtered through old high trees. He couldn't see the guards at the rear of the HQ house so that they couldn't see him, he figured. Only one soldier at the barn door now. He was leaning against the wall, two feet from the door, and four feet away from him. He was in shadows. He checked his watch, six minutes to go.

He stepped around the corner towards the guard while saying something in Korean, and plunged his knife into the guard's throat. Foxy held him so he wouldn't fall or make noise from the thrashing of his legs, and listened to the gurgling sound as the enemy soldier drowned in his own blood. He waited till all was silent. No alert. He dragged the body around the corner.

He was about to open the door when he heard a very faraway gunshot. Oh, he thought, I hope that's good.

He pulled open the door and quickly shut it. The enemy officers turned to see a .45 automatic pistol aimed at them by an American soldier. They also saw his Tommy gun slung on his shoulder.

The torturer officer jumped up from his seat at the table and yelled, "Who the.." that's as far as he got.

"Shut up! You say one more word, and I'll shoot you in the face. Sit down!"

He sat.

"Put your weapons on the table, any funny moves, and you're dead. NOW!"

The two complied, but he saw the one Korean officer hesitate until he watched the other men relinquish their weapons. He then drew his gun very slowly and tossed it in with the others. Foxy gathered the pistols and threw them into a corner.

He saw the Korean torturer was a major, and the other Korean was a captain. He also saw a unique insignia arm patch on both. Strange, he thought, a red world with a gray dragon on it.

"Been watching for a while. I know a gunshot in here doesn't bring shit, so everybody remains seated."

Foxy kept his .45 on them as he went over to the hanging soldier on the cross and saw that new things had been done to him. He was barely alive but still conscious. He went to his side so he could keep watching the men at the table and leaned close to his ear, "Hey buddy, I'm with Nelson, are you?"

In a faint whisper, as the man turned his head slightly towards him, he uttered. "Yes. Nolan…. Franks and Rici dead. Don't know about the rest."

Nolan then gave out with a slight exhale, and his head dropped down.

Foxy raised and saw the indifference on the three officers sitting at the table.

"Hey, assholes. Let me introduce myself."

A distant explosion resonated in the barn. Foxy didn't flinch, but they did. They looked at each other but remained silent. When three more explosions were heard and felt, Foxy smiled and relaxed, but the three officer's expressions got appreciably more apprehensive.

He kept grinning at them, seeing their anxiety. "Let me finish my introduction. I'm an American professional soldier. Like the three you just killed here." As he stared at them with pure hate in his eyes, he said, in a smooth and dark tone, "It's regrettable you chose to do these things to my fellow soldiers."

The House – 2302 – (11:02 pm)

Mario and Tracker waited on the north rear side of the house. The first far off explosion set their internal clock. Two guards rounded their corner at the HQ house and stopped when they heard the detonation. They spoke, decided to unsling their rifles, and resumed their perimeter walking at the ready. As they got close to the waiting men, the three sequenced explosions echoed through the valley. The guards stopped again, talked with anxiety when Mario and Tracker silently killed them.

Tracker removed the dead guard's helmet and put it on, picked up his rifle in his left hand, and motioned to Mario to follow him. They then moved to the rear of the house corner where he abruptly ran around the corner and yelling the only lines of Korean he had memorized from Foxy's language cheat sheet he had made so long ago, "Help! Help! Come quick!"

The guards were already on alert. The light was dim, and Tracker did well with his pronunciation. The two soldiers turned side by side to move to the corner—big mistake.

Mario and Tracker moved quickly to collect their knives, unslung their Tommy guns, and moved to the back door entrance to the HQ house.

Out front, the first bridge explosion roused those who were missed by Tracker and Mario's grid killing. They were tired, but they'd heard distant blasts before. But this was pretty big! So what, most thought and rolled over.

The next three detonations, however, rocked their world. Something serious just happened. The men that got up were somewhat disorganized because many of their comrades and leaders were not rising along with them.

Boom! Boom! Boom! The trucks in front of the house exploded within a few seconds of each other.

Mulling about and confused, these soldiers became easy targets for the exploding vehicles and the flying shrapnel they produced.

Mario was first through the HQ buildings' rear door. He fired on entering, spreading the clip around the entire room, hitting a man in the corner, then an oil lamp, then another man. Tracker jumped in beside him and threw a grenade down the left hallway, then dived into the south parlor.

The oil lamp spilled onto the floor, spreading flames.

Boom! A yell from down the hall where the grenade exploded. Mario moved to the corner of the right hallway. A pistol stuck out from a doorway, and shots fired wildly at him. Mario responded with a short burst, then a scream. Then something in Korean.

He moved quickly down to the door and saw a soldier writhing on the floor, holding his hand. He wanted to kill him but thought, is this the general? So he didn't. The light was dim; he could only see that this guy was out of action. He heard more shots towards the front and moved. He opened a second door farther up the hallway and saw a soldier at a desk.

The back room of the house was now fully engulfed in flames.

The Barn – 2305 – (11:05 pm)

The explosions and shots nearby sent a wave of foreboding through the officers at the table. They knew bad things were coming their way.

The Russian officer was first to speak. In a thick accent, he said, "I'm Colonel, only adviser to unite these two unfortunate separated peoples. I mean no harm to America."

Foxy walked up to him. "Colonel, were you just watching the show? You did nothing to stop this torture. Had I'd known you Russians were as bad as the Nazis, I wish they'd slaughtered you all in the last war."

"What means? I do nothing!"

"That's right! You didn't do anything you piece of shit! Do you believe in God?"

The Russian squinted his eyes and showed fear. "I believe in State, no God."

"Say a quick prayer to Mother Russia, comrade." Foxy raised his .45, pointed at the Russian colonel's head, and fired. The other two officers jumped, totally shocked. Foxy moved his .45 quickly to each of them. They saw death and got still.

In Korean, Foxy spoke to the quiet officer who reseated himself. "What's your role here? You don't look like an evil man."

The other Korean officer, the major, sitting on the other side of the dead Russian, spoke harshly to him in Korean "Remember your place! Remember our mission."

Foxy looked at this quiet guy again, then looked over at the big mouth major. He looked back at the quiet guy and saw a totally different look, pure hatred. He couldn't believe it.

Foxy spoke in Korean again, asking the captain, "What's your mission? And what is your unit?"

His eyes changed from hatred to confusion and looked over at his major.

Foxy stared at him and continued in Korean, "Captain, you're going to die here today. It's just a matter of how. Answer my questions, and I promise it will be swift."

The major started yelling at the captain. Foxy took two steps toward the major and hit him across his head with his .45, knocking him off his seat. He hit the floor and didn't move.

"Now, Captain!" Foxy resumed in Korean.

"Foxy!" Came a yell from Arron at the rear opening. "Tully coming through the front, don't shoot."

The barn door opened. "Tully here! Don't shoot!" He yelled in. The magnitude of the gunfire outside now registered on Foxy.

"Okay! Come!" Foxy yelled back, needing to raise his voice above the noise.

Tully entered low and went left. Joe followed and went right. What they saw was bizarre. An oil lamp at the table showed the interior barn. At the table was a dead guy slumped in his chair with a hole in his head. A man hanging in Tully's corner with no skin almost made Tully vomit. Backing away, Tully tripped over a dead guy with his face shot off.

"What the fuck?" Joe exclaimed as his first sight, was the naked dead man on the cross, in the middle of the barn, with missing body parts.

Foxy stood with his .45 aimed at a North Korean captain sitting in a chair. Another Korean officer lay nearby on the floor, hand on his bloodied face, moaning.

Foxy looked at Tully. "I'll explain later!"

Tully was quick to his senses. He had no idea what was going on but knew Foxy was in control. He didn't want to get killed by Mario or Tracker, nor did he want to kill them in this confusion. Firing around the main house had intensified.

"Where are they? What can we do?" Tully was intense.

Foxy had anticipated the situation and planned for it.

"They're in the house. They're expecting you to seal the north and south sides when you arrive. The rear should be clear. Do not go to the front. Anyone coming out the back is friendly unless they're Korean, so don't shoot first."

"Be back!" They exited the barn. "Joe! Take the south side with Moose," who was waiting just outside the barn. Tully yelled. "Moose, follow Joe."

The fire inside the rear of the house had gotten bigger, now moving up to the roof.

Tully ran at an angle towards the north corner of the house and turned his head, "Arron! With me!"

The light from the fire made them totally exposed as they crossed to the north corner of the house. Two enemy soldiers came running around the same corner, fifteen feet away. Tully reacted first.

Arron was stunned at how fast he moved. "You Rangers have to keep putting us mere mortals down regularly, don't you?" A pause. "Thanks!"

"Can't let a great cook die, can I? Move!" They rounded the house corner, dove to the ground, and started firing.

On the south side of the house, Joe ran just past the house corner and signaled Moose to come close to him. "There, under that bush, you take out anything that comes from that front corner. I'm going inside."

"But Joe! We're supposed to stay together, right?"

"Something bad going on inside! I gotta go! Nothing gets through you, right?"

Joe didn't wait for an answer; his instincts were on fire.

He stood, ran ten feet and propelled himself headfirst through a side window, glass and wood splattering, he screamed, "Joe here!" He crashed through, hit the floor and rolled to the far wall, with his Tommy at the ready.

A figure crouched in the front doorway started to turn his way, his weapon highlighted by the fire in the adjacent room. The man's head exploded into Joe's place.

"Tracker here!" Joe heard from the other room. "What asshole just saved my life?" Tracker yelled again.

Joe moved to the doorway and yelled in, "Not sure who saved who Tracker. Where's Mario?"

Joe entered the room where Tracker was behind an overturned sofa near the hallway door. Bodies lay about the floor.

"Mario's on the other side. It just got quiet over there. Two hallways with rooms between them, I think—one room on each side. We need to search. No general yet."

They heard an intense firefight going on from outside.

"Stay here! I'll go around and come up the hallway from the rear." Tracker nodded. Joe moved back to the room he had jumped into, then into the hallway. The rear room was burning fiercely now. The inferno raced along the ceiling as dense smoke filled the house.

The first door he found turned out to be a small closet. A waste of time, he thought, damn it, I need to move faster. Time was running out as the hall ceiling caught fire. At the second door, he went small and off to the corner, fired his Tommy into the overhead door towards the ceiling, and yelled in English, "Surrender or Die!"

"Fuck You!" Three shots came through the door.

"Have it your way!" He emptied his clip through the door. Opened it slightly, tossed in a grenade, and dove to the floor away from the blast area.

"You okay?" Tracker yelled from the front.

"Yeah! Find Mario before we get crispy!" Joe was getting up from the floor when the front door blew in. Wood splinters showered Joe down the hallway, knocked him to the floor. Tracker would've been dead had he not moved but a few seconds earlier.

Unhurt by the blast, Joe stayed prone, grabbed two grenades off his belt, and tossed them through the front door. Then he fired his Tommy through the opening. He turned to go to the rear, but that was a no go. The fire blocked that exit.

The noise was deafening. Smoke was so thick he could hardly see and now he started choking. It's getting really ugly, he thought.

The rear of the house collapsed.

The hallway ceiling came down around him, and a beam glanced off his shoulder. He started to crawl to the door. His mind became fuzzy. Is this it? No way I'm gonna burn. Got no choice, he thought.

As he neared the opening, a massive volley of automatic fire erupted from the front of the house. Then a series of short bursts. Joe vaguely heard shouts but was choking badly from the smoke and felt he was about to pass out. So he pulled out a grenade and said to himself he wasn't going to burn or be taken alive.

Almost blind, not sure where he was, he stumbled over a body at the front door opening and fell out and went through a three-foot hole in the porch. The grenade flew out of his hand. He hadn't pulled the pin.

"Who the hell is that!" Screamed Frenchy.

Rusty yelled. "Anyone in that house could be ours! Find out before you shoot!"

Rusty's team had returned to Red's position in time to see the trucks explode. As they watched the HQ drama unfold, Rusty didn't know Tully's specific plan but figured he would probably move from the rear to the front after creating some diversion in the front. There was no traffic on the road after they blew up the bridge, so all was clear from the north, but they had to be watchful for NK forces coming back up from the south that might respond to this attack. Red told him about the two entrance guards disappearing some time ago.

"Red. Stay here. You've got to cover our ass. I'm going in."

Rusty then led his team across the road at the run, down the entrance road, and into the front yard, where they immediately opened fire, catching the remaining enemy resistance from behind. It was a brief firefight.

As the fighting ended, Frenchy dragged Joe to a nearby tree and checked him for wounds beside the horrible slice across his forehead. Frenchy at first thought Joe was shot in the head when he pulled him out from the porch because of all the blood, but then realized what it was and wrapped his head in a big wound bandage.

Tracker and Arron tended to Mario, unconscious with a gash on his head and a wound in his side. They were patching him while Tracker was trying to stop the bleeding on his side.

Marty came over, "Jesus! That's a bad knife wound."

Tracker looked up, "Marty, I can't stop the bleeding."

Rusty came over. "How bad is it?"

Marty shook his head.

Tracker answered. "Mario may die."

It's then that Tracker got real still, made a few strange sounds, and looked up and called out, "Arron! Get that dead soldiers bayonet!" pointed. "Heat it in that fire by the truck, get it real hot! Hurry before Mario bleeds out!"

Arron sprang into action. Pulled out the big twelve-inch blade from the soldier's belt scabbard, grabbed his rifle lying next to him, ran to the fire as he clipped the bayonet on, and stuck it into the fire.

"You know what you're doing?" Rusty looked worried.

"Yes!" As Tracker watched Arron, "We didn't have doctors on the reservation. Hope this works."

Arron ran back, unclipped the red hot bayonet with a wrapped cloth, and handed it to Tracker.

Grabbing it, he held it still for a few seconds. Mumbled something, then slowly proceeded to insert it into Mario's wound, turning it slightly. He mumbled throughout this process. The sound of sizzling flesh was disturbing. The smell was worse.

Tracker finished. Then looked at Arron and Rusty. "He has a chance." Marty finished covering the wound.

Marty then went back to Joe.

"He was bleeding like a stuck pig." Said Frenchy, "I wrapped his head to stop it. It's a big gash!"

"Has he regained consciousness?"

"He's coughed a lot, but no."

"There's no other damage, think he's got a concussion, but he'll need assistance getting out of here."

When Marty left Mario and Tracker, Rusty knelt next to Tracker. "Do you know what happened in that house?"

"Not sure. Lots of dead guys, a few officers. I found Mario in a small room, a dead officer at a radio transmitter, and the general we were looking for, dead on the floor, shot in the face. The general had a knife in one hand, and the other hand was partly blown off, still bleeding."

"Was the radio working?"

"Jesus Rusty! The damn house was burning down. I don't know!"

"Sorry, Tracker. You did a great job! Then and now!"

Tully came up to Rusty, "I've got Moose and Dean getting a makeshift stretcher together for Mario. Larry's back on the railroad tracks setting more charges. Frenchy and Snake are taking Joe back to Red. Now you need to come with me to see Foxy."

As they moved to the barn, Rusty asked, "Is he okay?"

"Yes, but what's in the barn, it's disturbing."

They entered. Foxy wasn't there.

Chapter Forty Five

Monday - June 26, 1950   Day Two of the Invasion
Korea - Red's Team across the Road – 2340 – (11:40 pm)

Tracker and Rusty had led most of the team towards the Peak ten minutes ago. Foxy crossed the road a little later.

"You okay?" Red asked, noting distress on Foxy's face. He passed, not stopping.

"Yeah! Just want to get out of here," Foxy called back.

Red thought his answer odd.

Red and Ben waited for Tully and Larry.

Although the mission was a success, they were jumpy. It wasn't over yet. The house across the way was burning like a city in the middle of a forest, and Rusty thought that maybe they got a distress message off. Red felt, so what if they got off a message? I'm sure someone got a message. Hell, they probably heard all about this little shebang in Moscow by now.

"We got lights coming up from the south!" Ben called over. "Maybe two miles or so."

Red stood up, swung his glasses up, and looked down the road. "Halftrack, trucks behind it, can't tell how many, but more than we want to deal with."

To nobody, Red said aloud, "Come on, Tully! We got maybe ten minutes. Damn it, Larry! You always want to blow shit up."

Five minutes passed. Red was thinking about repositioning further south to hit the column before it got too close to the HQ house road entrance. He had one minute to make his decision.

"Red!" A yell, "Coming across!"

"Come fast! Company's coming!"

"Pack up, Ben!" Red removed the ammo belt from the gun, then lifted it out of its base and placed his machine gun over his shoulder. Ben loaded the ammo belts around his neck and stuffed others in his and Red's backpacks. They started moving as Tully and Larry came in with them. They all picked up their pace.

As they climbed higher and got further away from the road, Tully realized it had gotten much darker. He could barely see Ben's back only two feet away. The foliage obscured the glow from the burning house, but more importantly, there was no moon. The men were quiet.

Up ahead, it was slow going. Carrying Mario's stretcher up over the rocky path and ducking around or under outgrowing roots and tree limbs required each man to relay back to the next guy the approaching hazard. In the case of Marty, who was in front of Dean, who was the frontman on the stretcher, he waited until Dean got closer to a danger area, and, like a mother hen, he made sure he didn't stumble.

It wasn't long before Red's team caught up with the main force.

Distant explosions boomed, one after the other. It was Larry's barn and railroad track surprises.

Red called back. "Larry! You just love that shit, don't you?"

"It does make my heart sing, Red."

"Keep it up! I like the tune."

It started to rain.

Within a quarter-mile of the Peak, Larry, being the last man in the column, started finding his markers. He set his booby traps, placing a grenade either top or bottom, tying it in, then securing a thin metal wire around the pin and bringing it across the pathway and tying it to root or limb.

By the time he got to the base of the Peak cliff, only he, Red, and Tully waited to climb up.

"They all set, Larry?" Asked Red.

"The music selection is ready, maestro."

293

Chapter Forty Six

Tuesday – June 27, 1950,   Day Three of the Invasion
Korea - The Peak — 0200 – (2:00 am)

Marty had just finished cleaning Mario's wounds and started an IV drip. Tracker sat at his side.

"That was one beautiful job, Tracker. You saved his life. Getting the bayonet, damn that fast thinking. But what amazed me was how'd you know how deep to go in to cauterize the wound?"

"I didn't."

"Great guess then! Too deep and you would have caused more damage. Not deep enough, and he would've bled out."

"No guess," Tracker said in an even voice. "Spirit did it."

At first, Marty thought Tracker was pulling his usual Indian stuff. But Tracker wasn't smiling.

Marty said. "You really believe this?"

Tracker stared at Marty. "Yes. I have experienced Spirit in my life. I've seen its power and been blessed with its love and protection."

"Marty, I'm not lying! I have no recollection of helping Mario. I only remember kneeling over him, feeling helpless. Not knowing what to do, I started praying to Spirit. Then I heard myself telling Rusty he'd be okay."

"No, shit!"

Later.

Red, Tully, and Rusty settled into a corner near the entrance to the big cave. They took the opportunity to catch up on the night's mission. A heavy rain precluded any urgency from leaving.

Rusty started. "Marty says Mario is going to be okay; stitched up the knife wound. Said Tracker saved his life. He's got no concussion from the head gash, just a few stitches and be in pain for a while, but be able to function, Marty thinks at 80%."

"What the hell does that mean?" Red barked.

"No idea." Rusty shrugged. "Joe woke up and got 15 stitches and has a concussion. Marty gave him pills, and he's sleeping."

"You talk to Foxy yet?" Asked Tully.

"No. Wanted to talk it out with both of you first."

"What's this all about?" Red looked concerned at both men.

"I guess you haven't told him anything about the barn?" Rusty asked Tully.

"No."

"Somebody better tell me something!" Red snapped.

Tully briefed him on the high points.

"So, these were Captain Jordan's men?"

"Never met any of his men, but I knew their names." Rusty continued. "Tully. Tell Red what Foxy told you when you went back after the house fight was over."

"Foxy was pretty juiced up, wired and started telling me about this special unit, that they were all Intelligence soldiers but much more. The general was setting this house up as a major interrogation camp for ROK officers, political types, and American personnel, who were of particular interest. Americans were to be treated harshly."

Red flared. "Bastard Commies! Did he get a unit name? Anything about their chain of command?"

"They have a special patch." Tully showed him one. "The unit is called *Grey Dragon*. The North Korean captain didn't know who the general reported to, but suspected a Russian general was involved, and maybe a Chinese general as well."

"Why would he think that?"

"The Russian colonel with them often talked about his boss and the arguments he had with his asshole comrade Chinese general."

"You think this guy was lying?"

"No, Red."

"Did he die slowly?"

"Let's just say he could've died faster."

"How about this major? He inflicted the pain, right?"

"He died very slowly."

"Good!" Red whispered.

"When Foxy finished telling me all this, he kind of deflated, looked troubled, guilty almost."

Rusty nodded, "I think he feels ashamed about what he did. I've known him for a long time. Torture is so out of character for him. That's why I wanted to talk to you first because I'm not sure what to say to him. I'm disturbed that he lost control and stooped to their animal level. But also proud of the job he did on the mission."

Red didn't hesitate. "We've been through a lot of shit together, Rusty. We were lucky we never came across any Nazi's that had just brutally tortured American soldiers. Because if we had, we would have lost it just like Foxy. I know you, Rusty. You're the best damn leader and man I've ever met, but I think you'd do the same."

"I agree with Red." Tully added, "Plus, he got us good intelligence. Christ Rusty! You saw what those bastards did to our guys. I thought Foxy was too gentle."

Rusty looked over at Tully, paused, then laughed. Surprising both men. "Thanks, Tully."

Rusty stood, "Going to find Foxy. Then we'll report to Colonel Nelson. Red, find Larry, get him ready on the VLF. Oh, before that, see if you can get Arron to whip us up some coffee."

As Rusty walked out, Tully glared at Red. "What was funny?"

He found Foxy alone in one of the smaller caves.

"Sure is rare for a soldier in the field to appreciate heavy rain, isn't it?" Rusty said and sat down beside him.

"Yeah," Foxy muttered.

"We took a vote. Me, Red and Tully. Decided you are to be our lead interrogator, with one key proviso."

Foxy looked at him. "What?"

"You can't treat the prisoners so gently next time. Agree."

Foxy stared at him for a second. His eyes moistened. Then a small smile. "Thanks."

"No, Foxy. It is I who thank you. You alone planned that house and barn mission. It was brilliant. Tracker told me about what you heard under that window in the barn. The patience you showed for the greater good. Few men could've done that. Forget anything you think you did was bad. Remember only the good. You're a good man Foxy! Now let's see if Arron's got some sock coffee brewing someplace."

Chapter Forty Seven

Tuesday – June 27, 1950   Day Three of the Invasion
Korea - On the Eastern Bank of the Han River North of Seoul
on the Western Invasion Route
0230– (2:30 am)

The rain continued all night, giving George cover. He moved at a good pace, passing the ambush the ROKs had laid for the enemy platoon. He hadn't encountered anything since then and now thought he was in front of the North Korean onslaught for the first time.

His exact position still eluded him. What he did know was that he was near the lower Han River near Seoul. Since most of the explosions and gunfire were now happening slightly behind him, he guessed that he had passed south of the ROK defense line.

Moving along the riverbank, he tripped over something soft. It was a dead soldier. He searched around, found four bodies, all NKs. Off to the side, he found a small rubber boat with one pontoon deflated, shot through with holes. Must have been infiltrators trying to cross the river, he figured. Kimipo Airport is on the other side; nothing else was over there. It has to be close by. What? Maybe a few miles northwest of Seoul? Damn it; it's my break, he thought. Still dark enough to get across the river without being shot by friendlies. Then he reflected on these dead guys. Shit.

No choice, he concluded. I've got to cross this river at some point anyway, and this way looks a whole lot better than trying to get into Seoul without being shot and sneak across the Han River bridge. But then, I still have to cross this river. I'm sure all the bridges in Seoul, are or will have been blown by the time I get there.

He grabbed his knife and started cutting away the deflated side of the rubber inflatable.

Pushing the one-sided boat/pontoon into the river, he hopped on the float and started paddling.

Chapter Forty Eight

Tuesday – June 27, 1950,   Day Three of the Invasion
Japan - Eighth Army Headquarters –
Colonel Nelson's Office - 0300 – (3:00 am)

"Colonel! Wake up, Colonel!"

"What! What's wrong?" Nelson awoke from a deep sleep.

"Sorry, Colonel. Nothing's wrong. Got an incoming transmission from the Peak. We're on standby."

"Oh!" He sat up, started to clear his head.

"Thought you might need this." His sergeant handed him a cup of coffee.

"Thanks." He took two gulps and got up.

He stopped at the duty officer's desk. Asked the captain if any situation updates had come in since he had left for his cot at midnight. None, he said.

At the VLF station, Tech Sergeant Eddie Doyle was waiting. "Good morning, Colonel, we're all ready, confirmed identity, the transmission is good."

"Good. Okay, Eddie, confirm that we're all set."

Eddie started clicking away.

"Ready now. Confirm location. Proceed with report. Over. N."

Within a minute, Eddie was transcribing the response that Rusty was relating to Larry to code.

"Peak. Both bridges down. House and general down. Forty-plus special troops at house, down. Special unit called Grey Dragon. NK, Russian, and Chinese suspected command structure. Russian colonel killed. They have orders to kill Americans. Nolan, Franks, and Rici tortured to death at this location, witnessed. Heavy rain, so no pursuit. One wounded can travel. Plan to stay at Peak. Need rest. Over. R."

"Damn!" Said Nelson to Eddie, thinking of Captain Jordan's men. "Tortured! What kind of war is this going to be?"

"I'm sorry, Colonel." Eddie knew the names of all the Post soldiers.

"Send this." Nelson paused.

"Great job. Any hard evidence of Grey Dragon? NK in outskirts of Seoul. Proceed south along tower mountain line. Find area twenty miles *south* of Seoul in the east corridor. Try daily for update. Over. N."

A response moments later. Eddie translated. "NK officer hard interrogation only. Dead Americans proof of orders. Over. R"

They signed off.

Slowly Nelson got up from his chair and wandered back to his private office, closed his door. Dead. Not just dead... tortured to death. His mind spun, remembering recruiting each man. Jimmy Nolan, the skinny Irish guy, Norm Franks from St Louis, Ralph Rici from Manhattan. Why? What is this Grey Dragon unit? What about the rest of the men? Where's Captain Jordan?

He sat at his desk, wondering. Who should he ask about this secret unit? For that matter, who should he tell about what happened? Would Howie know anything? He doubted it. Would anybody believe it? Russians, Chinese, torture, orders to kill Americans? Would it matter?

They got those bridges! Hot damn! It was something positive! He focused on this impossible mission. It was his men, and they did it just like he knew they could. He thanked them and prayed for their safe return. But then his mood darkened again.

He stared at his calendar, then glanced over at his wife's picture, "Damn!" he said out loud, "I miss you, Ellen." I'm some lucky guy, he thought. That woman has put up with my career for eighteen years. Be honest, she's put up with you, period. Hard to believe she loves me as she does. But man! Do I love her! I'll call her later.

Korea - The Peak – 0330 – (3:30 am)

"What do you make of those orders?" Red said as he looked at Rusty apprehensively.

"I'd say it doesn't look good for the home team right now, and that's a long trip."

"Ya think?" Tully was angry. "Where the hell is that, fifty, sixty miles from here? You got nothing there, no contacts, no hideouts! Damn! You don't even know your way around. How are we gonna do this?"

"I'm glad you finally added *WE*! Because *WE* are going to do this." Rusty was stoic. "What happened to all your Ranger bluster?"

"Don't give me that horseshit, Rusty! Conducting high-risk missions is one thing. But this! Jesus! Maybe I'm losing it, but it just seems impossible!" He sat back; the wind was out of him.

Rusty knew they were all exhausted. Further discussion would achieve nothing. "Let's sleep; we'll talk later. Hopefully, it'll rain all day."

Japan – Eighth Army Headquarters – Colonel Nelson's Office - 0500 – (5:00 am)

His desk phone buzzed. "What's up, sergeant?"

"You better get in here, sir, all hell's breaking loose."

He slammed the phone down and headed for the conference room. Rounding the corner, he heard the phones ringing, the teletype machines clanking, and men shouting. What in hell! He thought.

Additional men were joining him as he entered the room, as they too had been roused to help out.

Cables were stacked at his table area as he sat when his Red "Kimipo Airport" phone light blinked. He picked up immediately. "Colonel Nelson."

"Glad I got you, Colonel." Captain Wild Bill Stands voice was very distinctive, "Billy over here just told me that he got a call from his buddy down at Suwon Airfield. They found a P-51 in a hanger that they're emptying of ammo and such for the ROKs. They wondered what they should do with it."

Wild Bill started talking fast now.

"Billy insisted I call you; he thought it would be a good idea for us, meaning Billy, me, and his sidekick to head down there and check it out. You know, him being an ace mechanic and me being an Ace pilot and all. Besides, we need to get the hell out of here because we hear artillery fire to the north of us as we speak. The air force guys that came in for the evacuation are gone, and only ROK troops are around. Most are manning the perimeter, filling sandbags, and such. And, oh yeah, we were strafed by NK fighters last night. Got us pretty good, too. Shot the shit out of a big Skymaster transport on the tarmac. Been burning like the great Chicago fire for the last nine hours."

"So what do ya say, Colonel! Can we get the hell out of here?"

"And good morning to you too, Captain! And why are you still on the phone with me. When you get there, you take possession of this plane as an Eighth Army asset. I'll issue the necessary orders after the fact. You're to tell no one about this, refer any inquiries to my attention. Got it?"

"You know me. I love working the edges! We'll be outta here quick as a rabbit."

"Good job on that rescue mission! Manny's gonna make it."

"I knew it! God *IS* on our side! Tally Ho, Colonel!"

The Red phone next to the one he just hung up started blinking, 'Korean General Staff G2 Seoul'.

"It's usually bad news when I get a call at five in the morning, Colonel Oh! Tell me it isn't so!"

"I can't. It just keeps getting worse, I'm afraid. We're moving our headquarters out of Seoul this morning. We've been fighting a delaying action in the northern suburbs since early last night. We'll fight in the city, block by block, but it is, as I've said, a delaying action. Later today or tomorrow, the city will fall."

"How many divisions do you have fighting there?"

"Divisions? No! No divisions! Battalions? Yes! Maybe a few."

"But Colonel, you had six of your best divisions north of Seoul, Over 50,000 men! Where are they?"

"It happened so fast. I don't know, a lot scattered, and a few are still fighting, but I think most are dead or captured."

"Where are you going to set up the new headquarters?"

"About a hundred miles south, in the city of Taejon. You know it?"

"Yes. It's a major relay station for the Korea/Japan communication cable. It's pretty far south, don't you think?"

"No! We don't believe that it's far enough unless America moves faster. Colonel, what I just told you about our force strength means we only have maybe three full divisions left, and right now, they are scattered. We haven't put a dent in the North Koreans. They keep rolling over us. So please, hurry up! We need your help!"

"We're moving! Have faith! Now I need your help."

"Anything!"

"My men from the Post, the men who just blew up those bridges at Cheorwon, are isolated in the mountains nearby. They need to get out of that area to be of any help to you. They can get to the northeast of Seoul, but no farther. They have no maps or local knowledge to go farther south. I want them to end up twenty or so miles south or southwest of Seoul. Can you help me with that?"

"I didn't know about the bridges! But I do know about your men, and they definitely are worth saving. Hold on a minute!" Nelson heard yelling in the background, some back and forth conversation. A few moments passed, then Colonel Oh was back.

"I think I have a solution. My nephew, Sergeant Chan-Jae, is with a signals unit stationed in that area. His outfit is in the hills just northeast of Seoul, well hidden, high up with a good view of the valley and Seoul. They are staying after we retreat to report on the enemy. Chan-Jae has family in Osan, a town twenty miles or so south of Seoul. I will order him to assist your men."

"Great! Is he capable?"

"Very. Came up the ranks through the Scouts, tough bunch. He's a physical kid with brains; speaks English too."

"Like it. Let's work out contact details." They did.

No sooner had he hung up when he heard, "Colonel, line one! General Walker."

"Is everybody mad this morning, General?"

"That's no way to greet a General, even if it is 5:30 in the morning." He grunted.

"Sorry, sir, it's just been a little crazy around here."

"Get used to it, Jim, because the asylum has just been taken over by the patients."

"What do you mean, General?"

"I take it you haven't read all your dispatches yet?"

"I've been on the phone since I got in. What's going on?"

"For starters, General MacArthur is flying into Seoul today."

"He what!?"

"It gets better."

"How could it?"

"Oh, you won't believe. The Seventh Fleet has been ordered into action."

"Well, at least that's a real positive."

"I didn't tell you where they were ordered to go, did I? I won't keep you guessing. A sane man would never come up with the right answer. They're moving into the Formosa Straight! The whole darn fleet!"

"Okay, now my head is spinning! I won't even guess as to why the fleet's going there, but there's more isn't there?"

"You bet! Major General Church is going with General MacArthur to set up his new Advanced Command Group (ADCOM) for all forces in Korea."

"Is he taking command of all Korean forces? Wait a minute! Has the President authorized sending in US troops?"

"The President has not given the go on committing American forces. He did, however, approve of our Naval and Air support below the 38th Parallel. As far as taking charge of South Korean forces, I don't know. But I think that's in the cards because they're meeting with all the big boys when they land, the Korean President, the Korean Army Chief of Staff and a few more."

"What's Eighth Army's involvement in all this, General? We are, after all, the muscle, right?"

"That's what I keep telling my staff. Plan for the worst, and keep planning, because soon, the call will come to do something. Right now, I got so many plans I'm running out of paper. I just got Colonel Harris starting up on plans for the defense of Pusan."

"Unless we do something pretty fast, General. At the rate of advance of the enemy and the potential total collapse of the South Korean Army, you may want to start planning for an invasion of the peninsula to recapture the whole of South Korea."

"That's next, Colonel! But I have faith in our President. He's not going to sit through this."

"Sure hope your right, sir!"

"Plan for the worst, Colonel! Do what you can to save these poor souls from themselves, will you? And remember what I said about big feet. Oh, have you heard anything from your Post soldiers?"

He brought him up to date on his men and hung up.

Nelson has had some wild days in my life, but this was turning out to be one that would reset the needle. Reflecting on what General Walker had just told him, about their response, about his planning and about what might or might not happen, he became somber. He had predicted it all and knew it was possible, very possible that the North Koreans could sweep the peninsula and declare the country unified, civil war over. Checkmate.

Korea - The Peak – 0900 – (9:00 am)

Rusty, Tully, and Red were sipping coffee in the big cave, ready to make their plans. They all had smiles.

"Amazing what five hours of good sleep will do for your outlook." Said Rusty.

"Great news from Colonel Nelson helps a lot." Said Red.

Tully was feeling a little foolish, "Okay, okay, feeling much better today."

"Good, I think we could still use more rest, stay until it stops raining or at least for one more day."

Red and Tully agreed.

"Let's get Arron, Tracker, and Snake to join us and go over the map book to plan our mission south."

Red went off to find the men.

"Sorry about last night, Rusty. That won't happen again."

Rusty looked at him. "Don't know what you're talking about, Tully. The only thing that happened last night was that we all went on one hell of a successful mission."

A little later that morning.

Joe was sitting in the far corner of one of the small caves sipping coffee. Dean wandered over and sat beside him. He pointed to the stitches in his forehead.

"You're gonna have a nice scar there, buddy, look like some fierce war hero."

"People will think I'm some kind of ex-con just released from prison."

"You are a hero, Joe. Don't talk like that."

"Do you think anybody will care?"

Dean took a sip of his coffee. "I've often thought of that." He got very still. "Joe, you and I have seen much and have saved many. We've seen copious amounts of blood on both sides, and yet we're still here, fighting in another war. Why? Have we become addicted to the high?"

"Maybe it's all of that, but from growing up in a tough neighborhood, it's always been about bullies. But don't go too deep on me, Dean. We were talking about my Frankenstein stitches." He paused, "And what the hell does copious mean?"

The End

Thank You for Reading the Post
The Journey Continues

If you enjoyed **The Post**, I hope you would continue to follow the series and all the characters into Book Two

Fighting Behind the Lines

www.ingramcontent.com/pod-product-compliance
Lightning Source LLC
Chambersburg PA
CBHW051239260626
47162CB00002B/512